Stars to Self: Chemical Cosmos

Mahima

Contents

Chapter 1

Introduction

Inherently curious, humans have long sought to understand our origin story. We enquire about our inception as a species, about the creation of the elements that comprise our bodies and the Earth, about the provenance of the Solar System, and about the connection between the the Milky Way (hereafter the Galaxy) and the Universe as a whole. This enterprise to define "our place in the Universe" has been approached via both philosophical pursuits and physical experimentation and exploration. I will stick to the discussion of astronomy's role as a tool in addressing some of these questions, and will omit my own metaphysical musings on these matters as a favour to the reader.

Observational astronomers are remarkably fortunate to have an entire Universe to use as a "laboratory" in the pursuit of understanding; however this freedom is not truly infinite. Constrained by the vast physical scales presented to us in the Universe, a very finite speed of light, current technology, and our existence as primates on the surface of a single planet, we need to develop clever methods to lever limited accessible information into a more comprehensive interpretation of the Universe. This book highlights how stellar spectroscopy can be used to illuminate our chemical origins, and presents some newly constructed tools used in this endeavour.

1.1 Stellar Spectroscopy

"Stellar spectroscopy" is an extremely expansive study that could describe a myriad of topics ranging from the solar corona to the integrated light of distant galaxies. In the context of this book, the study of stellar spectra will primarily be limited to the detailed chemical abundances of the oldest and most metal-poor[1][2] stars and their connection to Galactic evolution. The basics of how chemical abundances are determined from stellar spectra are presented in this section.

The light we collect from a star primarily reflects the outer, optically-thin layers of the stellar atmosphere. In these layers, photons produced by the underlying optically-thick black body are scattered and absorbed by free electrons, atoms, and molecules. Since these particles can only absorb photons of specific energies/wavelengths, energies dictated by the kinetics of any free electrons, the electronic structures of atoms, and/or the rotational and vibrational modes of molecules, each population of present particles will produce a unique spectral fingerprint. A majority of the work presented in this book uses spectra collected over optical wavelengths ($\sim 4000 - 8000$Å), where the majority of absorption lines are atomic transitions, i.e. electronic transitions between two specific energy levels in an element with a particular ionization state (number of electrons). In order to map an observed spectral line to a chemical abundance, one needs to determine the number of atoms present in a given ionization state and the number of electrons each atom has to offer. The Saha and Boltzmann equations encapsulate this information.

Under the assumption of Local Thermodynamic Equilibrium (LTE)[3], the number of atoms N in two ionization states i and $i + 1$ is given by the Saha Equation:

$$\frac{N_{i+1}}{N_i} = \frac{Z_{i+1}}{Z_i} \frac{2}{n_e h^3} (2\pi m_e k T_{eff})^{3/2} e^{-(E_{i+1}-E_i)/kT_{eff}} \tag{1.1}$$

Z are the partition functions describing the number of configurations for a given ionization state, n_e is the number of accessible electrons, m_e is the mass of the electron, E are the energies of the configurations, k is the Boltzmann constant, and T_{eff} is the effective temperature of the line forming region. Each atom in each ionization state also has a defined number of electrons in any given energy level. For given energy levels a and b, the number of electrons available N is described by the Boltzmann Equation:

$$\frac{N_b}{N_a} = \frac{g_b}{g_a} e^{-(E_b-E_a)/kT_{eff}} \tag{1.2}$$

---

[1]"Metals" refer to any/all elements heavier than helium

[2]Metal-poor stars in this book will often be classified by their metallicity [Fe/H], which represents the logarithmic ratio of Fe to H in a star, relative to the value measured in the Sun. Very Metal-Poor (VMP) stars are defined as having [Fe/H]$\leq$ -2.0, Extremely Metal-Poor (EMP) as [Fe/H]$\leq$ -3.0, and Ultra Metal-Poor (UMP) as [Fe/H]$\leq$ -4.0

[3]LTE assumes that the spatial scales on which temperature fluctuations occur are greater than the mean free path of the photons in the gas, and that the particles in the gas follow a Maxwell-Boltzmann distribution, which can be used to define a kinetic temperature of the region.

where g_a and g_b are statistical weights for the electron configurations. Combining the Saha and Boltzmann equations describes the relative number of atoms in different electronic configurations. These equations are frequently wrapped into stellar radiative transfer codes such as MOOG[4]. In conjunction with lab determined atomic information (statistical weights, partition functions, and energies) and a model stellar atmosphere (see Sec. 1.1.1), these radiative transfer codes can synthesize synthetic spectra with a defined chemical abundance for a species of interest. These model spectra are then compared to observed spectral lines to determine a chemical abundance for a given element and ionization state in the observed star.

1.1.1 Model Atmospheres and Stellar Parameters

Since absorption lines are only produced in the optically thin regions of a stellar atmosphere, only[5] these regions need to be modeled for radiative transfer codes. The model atmospheres used in this book are 1D plane parallel and 3D spherically-symmetric MARCS models (Gustafsson et al., 2008). The structure of these regions can be described by a few key thermodynamic quantities: temperature T, gas and photon pressures P_g and P_γ, gas density ρ, opacity τ, scattering coefficients κ, and number of accessible electrons N. This full suite of properties are often reduced to the "stellar parameters": effective temperature T_{eff}, surface gravity log g, metallicity, and microturbulent velocity ξ.

Effective temperature T_{eff}

The effective temperature of a star, T_{eff}, describes the temperature of the layer in the stellar atmosphere where the black body continuum forms. This temperature is wavelength dependent and often assumes LTE. T_{eff} can be determined spectroscopically by balancing the measured abundance from individual lines of a specific element (often iron) with the atomic excitation potential for each line transition. In other words, the slope of the linear regression to the measured abundances for each line vs. the excitation potentials for each line acts as a proxy for the atmosphere layer where the line forms. Therefore, the slope should be zero in a chemically homogeneous model atmosphere. This process is successful when there are a sufficient number of lines over a span of ~~excitation potentials. Species like~~ Fe I are often used due to the large number of Fe I spectral lines in the optical regime.

---

[4]MOOG was originally written by Chris Sneden (1973), and has been updated and maintained, with the current versions available

[5]The choice of the word "only" should be no means diminish the immense amount of computational work and scientific value in producing these atmospheres.

If the model atmosphere is too hot, lines produced at higher excitation potentials will be over represented (i.e. a high measured abundance) whereas a cold atmosphere will over represent the lines with small excitation potentials. Alternatively, empirical relations like the Infrared flux method (IRFM) can be used to determine T_{eff} when photometric colors and metallicities are known (Blackwell et al., 1979b; Alonso et al., 1999; Ramírez & Meléndez, 2005; Casagrande et al., 2010).

Surface gravity $\log g$

The surface gravity is defined as $\log g = \log(GM/R^2)$ where M is the stellar mass and R is the radius of the star up to the line forming region. The gravity of a star will affect the gas and electron pressures P_g and P_e in the line forming region. This also results in pressure broadening of spectral lines as van der Waals forces and dipole coupling become stronger. The electron pressure is also related to the number of electrons from particular ionization states; if we assume that the measured abundance for all ionization states of a given element should be equal (e.g. there is only one Fe abundance for an individual star), then we can derive $\log g$ spectroscopically by measuring the abundance of two ionization states of the same element. Fe I and Fe II are often used and the gravity of the model atmosphere is adjusted until A(Fe I) = A(Fe II)[6] All of the above assumes LTE. In the case of non-LTE (NLTE)[7], various ionization states, and even different absorption lines of a particular element, can be affected. Deviations in the derived chemical abundance between LTE and NLTE can be up to ± 1 dex (Mashonkina et al., 2017d), depending on the element and choice of lines. This book assumes LTE, however this assumption is investigated to some degree in Chapter 3 as non-LTE effects are expected and observed when determining chemical abundances, especially in RGB stars. Surface gravity may also be derived empirically if the distance to a star is known through a relationship between stellar mass, T_{eff}, and absolute magnitude (see following section on Bayesian inferred stellar parameters).

---

[6] $A(X) = \log[N(X)/N(H)] + 12$

[7] In NLTE, the velocities of the gas particles are still described by a Maxwell-Boltzmann distribution, however an external radiation field is also considered. Assuming the radiation field is not in equilibrium (does not follow the Planck distribution), it will affect the the populations of particles found in various energy levels, meaning that the kinetic equilibrium equations should be used over the Saha-Bolzmann equations. The rate at which inelastic/radiative transitions occur, in comparison to the rate of elastic collisions, will dictate the departures from LTE.

Microturbulent velocity ξ

Microturbulent velocity ξ refers to any and all small scale convection, which affects the gas velocities in the line forming region that are not accounted for ain a standard 1-D LTE analysis. The effect of ξ is displayed in spectral lines via a small Doppler broadening. Microturbulence has also been proposed to represent neglected NLTE corrections in the model atmospheres themselves. As a spectral line broadens, the measured equivalent width (EW) also increases, translating to a higher than expected measurement for the chemical abundance. ξ can be determined spectroscopically by requiring the abundances measured from multiple spectral lines of the same species to be independent of EW. ξ and $\log g$ have been found to scale together and this book uses the scaling functions described in Sitnova et al. (2016) and Mashonkina et al. (2017a).

Metallicity [M/H]

The total metal content of the stellar atmosphere dictates the number of atoms available for electronic transitions and the opacity of the region. Atomic information and preliminary abundances for elements like C, N, O, Fe, and the α-elements are commonly incorporated into stellar atmosphere models.

When denoted as [M/H], the metallicity represents the column density of all "metal" particles relative to the column density of H, scaled to the column densities observed in the Sun[8]. More generally, $[\text{X/H}] = \log(N_X/N_H)_* - \log(N_X/N_H)_\odot$, where N_{X_*} is the number density of element X in the observed star and $N_{X_\odot}$ is the number density of the same element observed in the Sun. Fe is frequently used as a proxy for M due to the large number of Fe lines in the optical regime. Metal-poor stars, like those studied in this book, are also commonly α-enhanced (with respect to the Sun) so the metal-poor and α-enhanced MARCS models have been used in the following chapters.

Bayesian Inference of Stellar Parameters

The spectroscopic methods used to determine stellar parameters, as described in the previous section, all require the presence of many (often > 100) spectral lines of a given species for statistically significant results. Unfortunately, VMP stars, and especially hot VMP stars, have few spectral lines in the optical regime. Consequently, these aforementioned methods are inappropriate or impossible to carry out. An alternative method used to determine

---

[8]This should not be confused with the mass fractions of hydrogen, helium, and metals: X, Y, and Z, respectively

stellar parameters, particularly for VMP stars, is to infer the stellar properties via statistical methods. Introduced in Sestito et al. (2019), this method, hereafter the "Bayesian inference method", utilizes photometric observables from *Gaia* (Gaia Collaboration et al., 2018) in conjunction with stellar evolutionary tracks to determine stellar parameters T_{eff} and $\log g$. The stellar parameters for the stars studied in Chapters 2 and 3 are primarily derived from the Bayesian inference method, so the mathematical formulation of this method, as shown in Sestito et al. (2019), is summarized, with permissions from the author, below.

In simplest terms, the idea behind this method is to use a theoretical stellar evolutionary track, an isochrone, to relate observable photometric colors and magnitudes to the stellar parameters T_{eff} and $\log g$. The notable challenge is the translation of observed apparent magnitude to absolute magnitude. The absolute magnitude of a star is dependent upon T_{eff}, $\log g$, metallicity, and stellar age, while the relationship between absolute and apparent magnitude is affected by the distance to the star and reddening.

Stellar distances can be determined geometrically from stellar parallax (ϖ), and with the high precision parallax measurements from the Gaia satellite, distances can be determined for the roughly one billion stars observed by *Gaia*. The lurking issue is that distant stars have unreliable ϖ, meaning a simple inversion of the parallax measurement to determine distance should be avoided (Bailer-Jones, 2015). Instead, one can combine photometric and astrometric properties with a prior describing the Galactic stellar density to create a probability distribution function (PDF) of the distance to a star. Given the observables Θ (*e.g.* photometry, metallicity, parallax) and a model $\mathcal{M}$, the posterior probability of a star being at a particular distance follows Bayes's rule:

$$\mathcal{P}(\mathcal{M}|\Theta) \propto \mathcal{L}(\Theta|\mathcal{M})\mathcal{P}(\mathcal{M}). \tag{1.3}$$

where $\mathcal{L}(\Theta|\mathcal{M})$ is the likelihood of the observables given the model, and $\mathcal{P}(\mathcal{M})$ is the prior describing the probability that the model represents the physical scenario.

Assuming the photometric and astrometric information from *Gaia* are independent, the observables Θ can be split into $\Theta_{\text{phot}} = \{G_0, BP_0, RP_0, \delta_G, \delta_{BP}, \delta_{RP}\}$ and $\Theta_{\text{astrom}} = \{\varpi, \delta_\varpi\}$, where δx the uncertainty associated with measurement x. In Sestito et al. (2019) the model is given as $\mathcal{M} = \{\mu = 5\log(r) - 5, A\}$, with μ the distance modulus of the star, r the distance to the star, and A its age. Expanding the above, equation 1.3 can be rewrttien as:

$$\mathcal{P}(\mathcal{M}|\Theta) \propto \mathcal{L}_{\text{phot}}(\Theta_{\text{phot}}|\mathcal{M})\mathcal{L}_{\text{astrom}}(\Theta_\varpi|\mathcal{M})\mathcal{P}(\mathcal{M}). \tag{1.4}$$

I'll refer the curious reader to Sestito et al. (2019) for the detailed derivation, but the key points are highlighted below.

$\mathcal{L}_{\text{phot}}(\Theta_{\text{phot}}|\mathcal{M})$ represents the photometric likelihood of a star for a choice of μ and A. In this case, MESA/MIST isochrones (Paxton et al., 2011; Choi et al., 2016; Dotter, 2016) are used to relate the choice of A to a set of predicted *Gaia* absolute magnitudes (M_G, M_{BP}, M_{RP}). These predicted magnitudes, can then be shifted by the choice distance modulus μ and compared to the observed photometric properties. This process is integrated along the isochrone to compute the highest probability μ and consequently, high probability distance. Two peaks in $\mathcal{L}_{\text{phot}}(\Theta_{\text{phot}}|\mathcal{M})$, are expected for most stars which correspond to the dwarf and giant solutions.

To break the degeneracy in the dwarf/giant solutions, the *Gaia* DR2 parallax ϖ and its uncertainty δ_ϖ are folded into the analysis. $\mathcal{L}_{\text{astrom}}(\Theta_\varpi|\mathcal{M})$ is defined as the normal distribution for ϖ given its uncertainty δ_ϖ. Even for stars with large δ_ϖ/ϖ, the *Gaia* data can often exclude the dwarf solution.

Equation 1.4 also requires prior information on the model $\mathcal{M}$. This can be decomposed into a prior on the heliocentric distance and the position on the sky in Galactic coordinates $\mathcal{P}(r|\ell, b)$, and a prior on the age $\mathcal{P}(A)$; $\mathcal{P}(\mathcal{M}) = \mathcal{P}(r|\ell, b)\mathcal{P}(A)$. The choice of prior on the line of sight distance is motivated by the known and expected distributions of old stars in the Galaxy and a combination of a thick disk and halo density profile (Binney & Tremaine, 2008; Hernitschek et al., 2016) are used. The remaining prior, $\mathcal{P}(A)$, may be the least certain as there is little constraint on the ages of the most metal-poor stars. With that said, they are assumed to be very old (Starkenburg et al., 2017b) and a uniform prior for ages between 11.2 and 14.1 Gyr is adopted (the maximum values of the MESA/MIST isochrone grid). The objective is then to infer the PDF on μ/the distance r to the star by marginalizing over the age:

$$P(\mu|\Theta) = \int P(\mathcal{M}|\Theta)dA, \qquad (1.5)$$

assuming $\mu \geq 0$ mag ($r \geq 10$ pc).

Finally, we can find the posterior probability as a function of $\log g$ and T_{eff} as each point on the theoretical isochrones corresponds to a value of the surface gravity and effective temperature. Figure 1 from Sestito et al. (2019) is reproduced below.

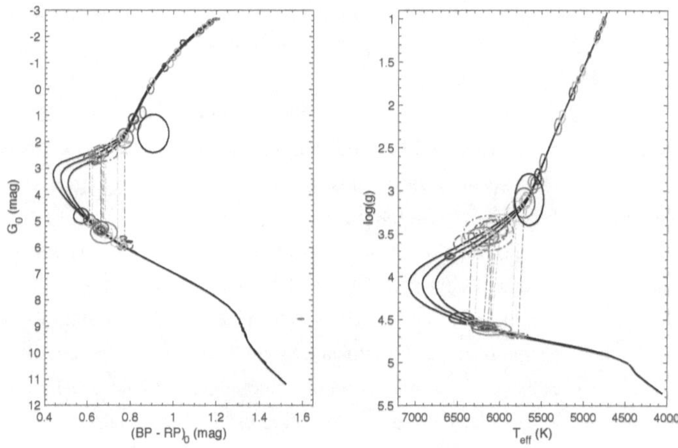

Figure 1.1 Figure 1 from Sestito et al. (2019). Position of their sample stars in the CMD (left) and the $\log(g)$ vs. T_{eff} plane (right). The ellipses represent the position of the stars within 1 sigma and the black lines correspond to the three MIST isochrones with $\log(A/\mathrm{yr}) = 10.05, 10.10, 10.15$ and metallicity [Fe/H] = -4.0 dex. If the dwarf-giant degeneracy is not broken, the two possible solutions are represented and connected by a dot-dashed line of the same colour code. Solutions with integrated probability ($\int_{d-3\sigma}^{d+3\sigma} P(r)dr$) lower than 5% are not shown and solutions with integrated probability in the range [5%, 50%] are shown with dot-dashed ellipses. Outliers in this figure, notably HE 033+0148, lay outside the colour range of the MIST/MESA isochrones.

1.1.2 Chemical Abundances

Prior to the formation of the first stars, the Universe was comprised solely of hydrogen, helium, and trace amounts of lithium and beryllium. Upon careful observation of one's own surroundings, a *very* keen observer may note that the chemistry of the Universe today is much more diverse. Assuming they are not too distracted by the dazzle of diamonds and the glitter of gold, they may ask themselves where all that C and Au came from. In the hot and dense crucibles of stars, the heavier elements are forged. Whether we consider the by-products of stellar evolution along the HR diagram (Herwig, 2005), the yields from core collapse supernovae (SNe) (e.g. Nomoto et al., 2006; Heger & Woosley, 2010; Tominaga et al., 2014; Ishigaki et al., 2018), or the by-products of neutron star mergers (Argast et al., 2004; Côté et al., 2017a), the diversity of formation sites is reflected in the diversity of chemistry of the Universe. By studying the detailed chemical abundances of individual elements visible in ancient stellar atmospheres, we can unlock clues about the formation of the elements themselves and the nature of the first stars. An overview of the principle groups of elements important for this book and their formation sites are given below:

Light elements The formation of the light elements (C, N, O, and some odd-Z elements like Na and Al) can be traced to a number of formation sites. He-burning in asymptotic giant branch (AGB) stars produce C and O (Herwig, 2005), the ejecta of Type II supernovae (SNe) and the stellar yields from the first stars show evidence of C and other light elements like Na and Sc (Umeda et al., 2006; Tominaga et al., 2014; Casey & Schlaufman, 2015; Clarkson et al., 2018), and convective mixing in massive ($M > 20M_\odot$) rotating low metallicity stars is a proposed site for primary nitrogen production (Meynet & Maeder, 2002; Hirschi, 2007). AGB stars play an important role in the enrichment of C in stellar atmospheres as well since convective dredge-up events bring C to the surface of the star where it may then be transferred to a binary companion or ejected into the interstellar medium (ISM) via stellar winds (Herwig, 2005; Arentsen et al., 2019). The CNO cycle (Burbidge et al., 1957) is the dominant H-burning process at temperatures above $\sim 2x10^7$K, where C, N, and O act as catalysts for the reaction (Kippenhahn & Weigert, 1994). As CNO cycling occurs deep in the H-burning layer of RGB and AGB stars, N is produced at the expense of C. Through consecutive convective dredge-up episodes, surface N is enhanced and C is diluted (Gratton et al., 2000; Spite et al., 2005). A study of the [C/N] ratio can provide insight to a star's progress along the giant branch and this observation has been used to estimate ages for evolved stars (Masseron & Gilmore, 2015; Martig et al., 2016).

α-elements: The α-elements are aptly names as they are formed via the capture of α-particles, ^{4}He nuclei, on a parent species. Associated with later burning phases in massive stars ($> 8M_\odot$) and SNe Type II, these elements are central in the study of the chemical evolution of galaxies (see further discussion in section 1.1.3). Typically Mg, Si, S, Ca, and Ti represent the classic α-elements, though Zn also behaves like an α-element in metal-poor stars and in damped Lyman-alpha systems (Kobayashi et al., 2006; Rafelski et al., 2012; Nomoto et al., 2013; Barbuy et al., 2015; Berg et al., 2015; Bensby et al., 2017; Kobayashi et al., 2020)

Iron-peak elements: The Fe-peak elements describe Cr, Mn, Co, Ni, and Cu. Capable of being formed in both SNe Type Ia and Type II, the primary contribution of these elements in galactic chemical evolution appears to be from Type Ia (Iwamoto & Saio, 1999; Tominaga et al., 2014). As sufficient time is needed to form the white dwarf progenitors of SNe Ia, these elements reflect time scales in galactic chemical evolution $\gtrsim 1$ Gyr after periods of star formation.

Neutron-capture elements: Nuclear fusion in stars is incapable of producing elements beyond the Fe-peak, meaning all heavier elements up to uranium need to be produced by alternative means. In regions with high neutron densities, light, stable nuclei will capture neutrons. These heavy isotopes are usually unstable and will decay via β- or β+ emission, ultimately changing the atomic number Z of the parent element. A majority of the neutron capture events will result in β- emission, which increases the atomic number of the daughter nuclei and creates a new element. There are two primary processes which act as pathways to the formation of the heavier elements: the slow-process (s) and the rapid-process (r).

Elements that are created when the neutron capture rate is slower than the typical β- decay time scale are the s-process elements. AGB stars are the primary source for the s-process and produce approximately half of the elements heavier than iron (Arlandini et al., 1999; Herwig, 2005). The third dredge-up event coupled with the strong stellar winds of AGB stars is largely responsible for the enrichment of the ISM with s-process material, Alternatively, when the neutron captures occur on time scale much faster than the decay time, the r-process, higher atomic numbers can be reached. Core collapse supernovae and neutron star mergers are currently thought to be the primary production sites of the r-process, though the diversity of r-process abundances in different environments and over a range of evolutionary scales suggest multiple pathways (Woosley et al., 1994; Travaglio et al., 2004; Freiburghaus et al., 1999; Tsujimoto & Shigeyama, 2014; Ji et al., 2016b; Côté et al., 2017b,a; Abbott et al., 2017; Drout et al., 2017; Roederer, 2017; Roederer et al., 2018a; Grichener & Soker, 2019; Siegel et al., 2019; Placco et al., 2020)

The contributions to an elemental abundance from the s and r-processes are often studied, as both processes often contribute to the production of a particular element. Eu is frequently considered a pure $r-$process element as the Eu abundance in solar-metallicity stars is explained by 95% r-process contributions and <5% of s-process contributions (Burris et al., 2000; Sneden et al., 2008). Alternatively, elements like Sr, Y, Zr, and La show over-abundances, relative to solar, at low metallicity. This suggest the nucelosythetic sites to produce these elements were different in the low metallicity Universe (Burris et al., 2000; Travaglio et al., 2004; Venn et al., 2004; François et al., 2007; Sneden et al., 2008). Galactic chemical evolution models can constrain the formation sites for the neutron-capture elements, but detailed chemical abundances for large samples of stars spanning the full metallicity distribution function are needed to test and constrain the theoretical predictions (Cohen et al., 2004; Norris et al., 2007; Heger & Woosley, 2010; Nomoto et al., 2013; Tominaga et al., 2014; Choplin et al., 2017; Côté et al., 2017a)

High-resolution spectroscopy (HRS) ($R \gtrsim 20,000$) is typically required to determine the detailed chemical abundance profiles of stars with high precision ($\delta < 0.2$ dex). In the case of detecting Eu in the optical regime, equivalent widths can be less than ~40mÅ in red giants, requiring $S/N \gtrsim 80$ and $R \gtrsim 40,000$ to detect. Spectral resolution R is defined here as $\lambda/\delta\lambda$ where λ is the wavelength of light of interest and $\delta\lambda$ is the smallest resolvable spectral element at that wavelength. $\delta\lambda$ is set by the width of the slit in long slit spectroscopy, the spacing of the gratings and/or the geometry of the grating itself in spectrographs that use a diffraction grating (e.g. echelle spectrometers like the Echelle SpectroPolarimetric Device for the Observation of Stars (ESPaDOnS) at the Canada-France-Hawaii Telescope (CFHT)), or the width of the mirrors used in an image-slicer (e.g. Integral Field Units like the Near Infra-Red Integral Field Spectrometer (NIFS) at Gemini-N).

With this said, HRS has its limitations. Bright stars and/or long exposure times are required in HRS to obtain a useful signal-to-noise ratio (S/N or SNR) as the light of the source is dispersed over many resolution elements (i.e. ~160,000 pixels in the case of a GRACES spectrum). Consequently, HRS studies have traditionally been limited to the solar neighbourhood or to the brightest sources in the Galactic halo and the dwarf galaxy satellites of the Milky Way. These limitations become prohibitive in building large HRS samples of distant/faint stars in the MW and its satellites.

A possible solution to the shortcomings of a high-resolution survey is to construct a medium-resolution spectroscopic (MRS) study. MRS permits the derivation of fundamental stellar parameters, metallicities, [α/Fe] , and *some* light element abundances with comparable accuracy, though lower precision, of measurements from HRS (Kirby et al., 2008). Since spectrographs with high spectral resolution require significantly longer exposure times than a medium-resolution instrument to reach the same desired S/N, medium-resolution surveys are typically capable of observing more targets, over a larger range of magnitudes, in a shorter period of time. These benefits make MRS the ideal tool to collect large samples of spectra for individual stars at large distances, enabling studies on the chemical evolution and hierarchical structure formation of Galactic satellites and nearby Galaxies.

Medium-resolution spectroscopic surveys like RAVE (R~7,000) (Steinmetz et al., 2020b,a) SEGUE (R~2,000) (Yanny et al., 2009), LAMOST (R~1,800) (Luo, 2015), and Gaia Radial Velocity Spectrometer (RVS) (R~11,500) Recio-Blanco (2016) have proven to be exceptionally valuable for studies of structure within the Galaxy. From medium-resolution spectra of $> 10^6$ stars, the metallicity gradient of the Galaxy has been explored (Schönrich & Binney, 2009; Grand et al., 2015; Kawata et al., 2017), age-metallicity relations have been identified and linked to Galactic kinematic history (Martig et al., 2014; Aumer et al., 2016; Grand et al., 2016; Casey et al., 2017), and the metallicity-distribution function of the Galaxy has been mapped (Casagrande et al., 2011; Hayden et al., 2015). While detailed chemical abundances are historically inaccessible from medium-resolution spectra as a consequence of spectral feature blending at medium-resolution, the derivation of fundamental stellar parameters is still crucial for Galactic astrophysics. Studies like Ting et al. (2017a,b) are challenging these historical shortcomings with machine learning methods.

The analysis of medium-resolution spectra can be most accurately treated via spectral modelling, similar to that done in high-resolution analyses. Given a set of stellar parameters and a well vetted linelist, a synthetic spectrum can be synthesized and compared to the observed spectrum. Though weak lines and line blends cannot typically be resolved in a medium-resolution spectrum, if the spectrum has sufficient spectral coverage (on the order of a few thousand Angstroms), then the spectrum will contain the integrated information of hundreds of absorption lines. This high information density generates strong statistical power when comparing the synthesized spectrum to the observed spectrum, allowing one to converge on optimal atmospheric parameters and abundances. This technique is the fundamental process behind the FERRE code (Allende Prieto et al., 2006) (see Chapters 2 and 3). The shortcoming of this process is that the individual effects of stellar parameters, metallicity, and individual abundances becomes obfuscated and correlations can arise. Careful selection of spectral coverage, linelists, and analysis methods is paramount in making reliable measurements from MRS. Chapter 4 highlights two projects related to improving the fidelity of MRS studies.

1.1.3 Chemical Evolution

The first stars that formed in the universe, the Population III (Pop III) stars, formed from gas composed solely of Big Bang nucleosynthesis products, i.e. hydrogen, helium, and trace amounts of lithium and beryllium (Steigman, 2007; Cyburt et al., 2016). These stars evolved, formed heavier elements through the aforementioned nucleosynthetic pathways, and dispersed these new elements into their environments through supernovae and stellar winds. Regardless of the formation mechanism, newly created elements are inevitably injected into the interstellar medium, polluting the pristine, post-Big Bang gas with trace amounts of metals. This seeded material may go on to form a new generation of stars with the chemical signatures of the earlier generation locked into their atmospheres. Subsequent generations of stars formed from this "enriched" material and continued the process of metal enrichment of the Universe, producing new generations of stars which are increasingly metal-rich. The Universe however does not have a homogeneous chemical distribution because the distribution of stars is inhomogeneous. Most stars are locked into structures like galaxies and dwarf galaxies which themselves have finer stellar substructures which may experience their own evolutionary paths. As a result, each stellar substructure may have a chemical fingerprint as unique as the next depending on structure characteristics such as stellar mass, dark matter halo mass, luminosity, the assumed initial mass function of the stellar population, star formation history, and star formation efficiency (Freeman & Bland-Hawthorn, 2002; Tolstoy et al., 2009). Thus, detailed chemical abundance maps of these systems, particularly for elements sensitive to well defined pathways, allow for us to not only probe the astrophysical processes responsible for the formation of the elements themselves, but the formation processes of the structures as a whole.

Chemical abundances have been measured for hundreds of very metal-poor stars in the Galactic halo and in nearby dwarf galaxies (Suda et al., 2017; Frebel & Norris, 2015; Tolstoy et al., 2009). It has been observed in the Milky Way and the dwarf galaxy satellites that

SNe Type II contributions fix [α/Fe]~ +0.4 (McWilliam et al., 1995; Gratton et al., 2000; Tolstoy et al., 2009; Mashonkina et al., 2017d). On larger time scales, after ~1 Gyr, lower mass stars will evolve into white dwarfs and SNe Type Ia will begin to contribute to the environment. Since the SNe Type Ia primarily produce Fe (and Fe-peak element) but trace amounts of α-elements, the [α/Fe] ratio decreases with [Fe/H] with the addition of Type Ia products, producing a "knee" in a plot of [α/Fe] vs. [Fe/H]. The location of this "knee" is sensitive to both the timescales on which SNe Type Ia become dominant over Type II, and to the star formation history and efficiency of a system. Dwarf galaxies have lower star formation efficiencies and shorter star formation histories ($< 1 Gyr$ Tolstoy et al., 2009) than larger systems, due to their lower gas and dark matter halo masses. Coupled with a "top-light" initial mass functions (IMF) (i.e., few contributions from stars over 20M$_\odot$), as suggested by Tolstoy et al. (2003); Hasselquist et al. (2017), SNe Type Ia occur at earlier times than in the higher mass systems (Salvadori & Ferrara, 2009). The end result is that the [α/Fe] vs. [Fe/H] "knee" in dwarf galaxies turns over at lower metallicities than observed in the Milky Way, but the lack of SNe Type II at later times permits [α/Fe]< 0 (Venn et al., 2004; Tolstoy et al., 2009; Frebel et al., 2010; Letarte et al., 2010; Nissen & Schuster, 2010; de Boer et al., 2012; McConnachie, 2012; Venn et al., 2012; Nomoto et al., 2013; Frebel et al., 2014; Hawkins et al., 2014; Berg et al., 2015; Hawkins et al., 2015; Venn et al., 2017; Hayes et al., 2018; Lucchesi et al., 2020; Nidever et al., 2020; Silva et al., 2020). The consequences of the different evolutionary histories of dwarf galaxies and larger systems is not strictly displayed in the α-element abundances. The diversity seen in r-process abundance ratios, in the Galaxy as well as in dwarf galaxy satellites, may be telling of the chemical evolution history of a system as well (Venn et al., 2004; Tolstoy et al., 2009; Venn et al., 2012; Ji et al., 2016a; Hansen et al., 2017; Roederer et al., 2018a; Hansen et al., 2018; Marshall et al., 2019; Sakari et al., 2019; Cain et al., 2020; Ezzeddine et al., 2020; Holmbeck et al., 2020b,a; Placco et al., 2020; Yuan et al., 2020). In the smallest dwarf galaxies with measured r-process abundances (e.g. Ret II and its analogs), the stars appear highly enhanced with r-process material (Ji et al., 2016a; Roederer et al., 2016). These low mass systems have less total gas to dilute any r-processed material that is produced. Alternatively, more massive systems like Tuc III appear to host a more diluted r-process signature (Hansen et al., 2017). Hydrodynamical models of dwarf galaxies also suggest that the *location* of the site of the r-process within a dwarf system can play a role in the abundance profiles seen (Safarzadeh et al., 2019; Tarumi et al., 2020). Studying the detailed chemical abundances of stars is clearly a powerful tool in understanding the evolution of a system.

1.1.4 Galactic Archaeology and Chemical Tagging

Knowing that the physical properties of a galaxy and its' evolutionary history can dramatically affect the abundance ratios displayed in its' stars, one can examine the chemo-dynamics of stars, the union of stellar dynamics and chemical-cartography, within the Milky Way to guide our understanding of the physical conditions in the early Universe, to explore epochs of early star formation, to identify substructures related to the dwarf galaxy progenitors of the Galaxy, and to test predictions from models of stellar nucleosynthesis, supernovae explosions, and galactic chemical evolution. Collectively, this endeavor is referred to as chemical tagging, Galactic archaeology, and/or near-field cosmology (Freeman & Bland-Hawthorn, 2002; Venn et al., 2004; Tolstoy et al., 2009; Frebel & Norris, 2015).

The oldest and most metal-poor (MP) stars are fossils of the early Universe. Observable today in the Galaxy and in its dwarf galaxy satellites, these relics are an indispensable tool for Galactic archaeologists to locally study the physical processes of the high redshift universe (Freeman & Bland-Hawthorn 2002; Beers & Christlieb 2005; Frebel & Norris 2015; Hartwig et al. 2018; Salvadori et al. 2019). Dedicated surveys such as the HK survey and Hamburg-ESO surveys Beers et al. (1992); Christlieb et al. (2002b); Beers & Christlieb (2005), the SDSS SEGUE, BOSS, and APOGEE surveys (Yanny et al., 2009; Eisenstein et al., 2011; Majewski & SDSS-III/APOGEE Collaboration, 2014), LAMOST (Cui et al., 2012), SkyMapper (Keller et al., 2007), and the *Pristine* survey (Starkenburg et al., 2017a) have uncovered the bulk majority of the known metal-poor stars. Though many EMP stars have been discovered (according to the SAGA data base (Suda et al., 2017), there are ~ 500 stars with [Fe/H]< -3.0 known), less than half have the detailed chemical abundance analyses necessary to place these objects in a greater cosmic context. Furthermore, after nearly two decades of searching for these rare stars, only $\sim$20 stars with [Fe/H]< -4.5 and only 8 with [Fe/H]< -5.0 are known (e.g. Aguado et al., 2017a, 2018a,b; Starkenburg et al., 2018; Bonifacio et al., 2018; Frebel et al., 2019; Nordlander et al., 2019).

EMP stars are interesting because they have only been enriched by one (or a few) supernova, and at ancient times before significant chemical evolution occurred in the universe. As mentioned in the previous section, Pop III stars formed from pristine gas. With no metals present to efficiently cool the gas via metal-line emission, large Jeans masses, and consequently massive stars ($M_* \gtrsim 100 M_\odot$), are expected (Silk, 1983; Tegmark et al., 1997; Abel et al., 2000; Bromm et al., 2002; Yoshida et al., 2006)[9]. These massive stars would have been short lived, quickly enriching their local environment with metal-enriched supernova

---

[9]This paradigm is being challenged with improved gas fragmentation models and the discovery of very low mass ultra metal-poor stars (Clark et al., 2011a; Schneider et al., 2012; Schlaufman et al., 2018)

ejecta. Depending on the intensity of the radiative feedback from the Pop III stars, and the mass of the dark matter mini-halos in which the first stars are expected to form (Tegmark et al., 1997), the metal-enriched gas may re-collapse and form a new generation of stars (Population II/Pop II) (Cooke & Madau, 2014). The presence of metals in the gas enables more efficient atomic line cooling and dust formation, facilitating gas fragmentation, and ultimately the formation of lower mass stars (Clark et al., 2011a; Wise et al., 2012; Schneider et al., 2012). If a $\sim 0.8 M_\odot$ star were to form from this polluted material, it could exist today on the main sequence, illuminating much earlier times. The detailed chemical abundances of all EMP stars are a unique measure to disentangle the effects of nucleosynthesis and galactic chemical evolution.

Where we find metal-poor stars today is also diagnostic of early galaxy formation. Based on cosmological simulations of the Local Group, it is thought that the Galactic halo was formed through the accretion and disruption of dwarf galaxies at early epochs. Consequently, the old, metal-poor stars seen in the halo manifest the properties of their progenitor systems (Ibata et al., 1994; Helmi et al., 1999; Starkenburg et al., 2017b; El-Badry et al., 2018a; Safarzadeh et al., 2019; Das et al., 2020; Tarumi et al., 2020). The arrival of precision proper motions from *Gaia* DR2 (Gaia Collaboration et al., 2018), in conjunction with increasingly large datasets of stars with spectroscopic radial velocities (RVs), has enabled the determination of orbits for halo stars. The majority of the UMP halo stars have been shown to have high-velocities and eccentric orbits, consistent with those expected from an accreted dwarf galaxy (Sestito et al., 2019). Similarly, a increasingly large population of halo stars has been found with highly retrograde orbits and kinematics consistent with the proposed halo merger remnants *Gaia-Enceladus* (Belokurov et al., 2018; Haywood et al., 2018; Helmi et al., 2018; Myeong et al., 2018; Monty et al., 2020) and *Gaia-Sequoia* (Barbá et al., 2019; Myeong et al., 2019). Curiously, two metal-poor stars have been found on nearly circular orbits in the Galactic plane (Caffau et al., 2012a; Sestito et al., 2019; Schlaufman et al., 2018). Since the Galactic plane is thought to have formed ~ 10 Gyr ago (Casagrande et al., 2016a), these stars challenge the idea that the most metal-poor stars are also the oldest stars. The Galactic halo is not the only place to look for the oldest stars, as Galaxy formation simulations predict that the Galactic bulge is another prime location to look for these relics (White & Springel, 2000; Starkenburg et al., 2017b). To date, only a small number of very metal-poor stars ([Fe/H] < -2.0) associated with the bulge have been found (Howes et al., 2016; Lamb et al., 2017; Lucey et al., 2019; Arentsen et al., 2020b; Lucey et al., 2020). Detailed chemical abundance analyses for these objects are limited, but they indicate many of the metal-poor bulge candidates are chemically similar to halo stars.

In fact, estimates of their orbits suggest these stars are likely just bulge interlopers; normal halo stars with plunging orbits. *Gaia* DR2 proper motions are paramount to find metal-poor bulge members (Lucey et al., 2020). Regardless of where we find metal-poor stars, their connection to local dwarf galaxy satellites and the formation of the Galaxy warrants their detailed study.

Though EMP stars are incredibly valuable tracers of the formation history of the Galaxy, the diversity of chemical abundance profiles and dynamics seen in this sparse population of objects makes for challenging statistical studies. Galactic archaeology will progress as statistically large spectroscopic samples of metal-poor stars are found in a variety of environments within the Local Group. Any and all contributions to the discovery of new EMP/UMP stars, and the spectroscopic follow-up of these rare and enlightening objects is valuable to this field. The following section introduces one such effort, the *Pristine* survey.

1.2   The *Pristine* Survey

The Pristine Survey (Starkenburg et al., 2017a) is a narrow-band, photometric survey focused around the metallicity sensitive Ca II H & K absorption lines conducted with MegaCam at the 3.6-meter Canada-France-Hawaii Telescope (CFHT). Photometric metallicities are derived though an empirical relationship which compares the flux from the CaHK filter to SDSS colours. Early comparisons with SDSS SEGUE metallicities for a subset of overlapping targets showed that metal-poor stars could be identified with a careful selection of colours (Starkenburg et al., 2017a). The most metal-poor candidates selected from photometry are then targeted with medium resolution ($R \sim 3500$) optical spectroscopy, primarily at the 2.5-meter Isaac Newton Telescope (INT) and 4.2-meter William Herschel Telescope (WHT), with spectroscopic metallicities and carbonicities derived from the FERRE code (Allende Prieto et al., 2006). *Pristine* has been remarkably efficient with success rates of 56% for finding stars with [Fe/H]< -2.5 and 23% for stars with [Fe/H]< -3.0 (Youakim et al., 2017; Aguado et al., 2019b). As of September 2019, *Pristine* has discovered 707 new VMP stars with [Fe/H]< -2.0, and 95 new EMP stars with [Fe/H]< -3.0 (Aguado et al., 2019b). Within the population of EMP stars discovered by *Pristine* is Pristine 221.8781+9.7844, the second most metal-poor star known in terms of total metals measured (Starkenburg et al., 2018). Through *Pristine*, it is expected to find one star with [Fe/H]≤ -4.0 for every ~ 100 stars with [Fe/H]≤ -3.0 or roughly ~ 15 UMP stars over the ~ 1000 deg^2 footprint, vastly increasing the sample of known stars with [Fe/H]≤ -4.0 (Youakim et al., 2017).

1.2.1 Observing Campaigns

The success of the *Pristine* survey can be attributed to the performance of the survey's photometric metallicity estimations, coupled with the target selection for the spectroscopic follow-up programme. When carrying out spectroscopic observations, the most interesting targets must be identified from the thousands of objects with estimated photometric metallicities to optimize the use of telescope time and to maximize the potential scientific return. In the current structure of the *Pristine* programme, this responsibility of target selection is left to the telescope operator.

The medium-resolution spectroscopic campaign for *Pristine* has largely been fulfilled at the INT (145 of 182 total nights of spectroscopic observations) using the blue sensitive EEV10 detector on the $R \sim 3500$ Intermediate Dispersion Spectrograph (IDS). Through two observing runs at the INT, spanning a total of 15 nights in May 2016 and May 2017, I selected and observed 110 and 117 targets, respectively. The 110 stars observed in May 2016 comprise 53% of the 210 star sample used by Youakim et al. (2017) to perform the first refinement of the *Pristine* selection criteria. The full sample of 227 stars that I observed comprise 23% of the total number of *Pristine* stars with follow-up medium-resolution spectroscopy (MRS). The MRS sample has been used by Aguado et al. (2019b) to further characterize the *Pristine* selection criteria and success rates, by Youakim et al. (2020) to probe the Galactic metallicity distribution function down to [Fe/H]~ -3.5, and by Sestito et al. (2020b) to examine a population of metal-poor stars in the Galactic disk.

The ~ 100 new EMP stars discovered by *Pristine* are strong candidates for high-resolution spectroscopic follow-up observations. In the 2018A observing semester, I led a Gemini/-GRACES campaign as PI to observe five EMP stars discovered during the the *Pristine* MRS follow-up program. GRACES (Chene et al., 2014) is the fiber-feed from the Gemini-North telescope to the ESPaDOnS spectrograph (Donati et al., 2006) located at CFHT. Capitalizing on the high efficiency of the 8-meter Gemini-N telescope to feed the high-resolution ($R \sim 60,000$), broad spectral coverage ($\sim 4000 - 10000$ Å) ESPaDOnS, GRACES is a well suited instrument for detailed stellar spectroscopic studies. This program was awarded 9.3 hours of Band A time and was run to completion. During that semester, my advisor Kim Venn, also requested, and was awarded, ~ 150 hours, over 5 semesters, of Gemini/GRACES time as part of a Large and Long Program (LLP) starting in 2018B. Since then, I have taken over a large majority of the LLP duties, selecting high priority targets from the *Pristine* MRS sample, preparing the observing and instrument configurations required, and writing a data reduction suite to process the GRACES data. The LLP data has the potential to increase the number of known EMP stars with detailed abundance analyses by 30%, a pivotal leap in sample size for such important objects. Furthermore, such a large sample of EMP stars increases the likelihood of finding the next "most metal-poor star" and/or new metal-poor stars with unique chemical signatures, providing critical and new constraints on the characterization of the first stars and first supernovae. The bulk of my book will be structured around the HRS analysis of newly discovered metal-poor stars found in the *Pristine* survey.

1.3   book Outline

This book contains a collection of projects related to stellar spectroscopy. In Chap-ters 2 and 3, I continue with the theme of chemo-dynamics of Galacic metal-poor stars and present two *Pristine* survey projects. Changing gears, in Chapter 4, I present two projects unrelated to *Pristine*: an exploratory expansion of StarNet, a deep learning architecture applied to the analysis of stellar spectra, to analyse medium-resolution IR spectra (Section 4.1), and my internship at the Gemini-N telescope developing a data reduction pipeline for the Gemini-N Near-Infrared Integral Field Spectrometer (NIFS) as part of the New Technologies for Canadian Observatories (NTCO) program (section 4.2). Chapter 5 contains some commentary on ongoing projects.

Chapter 2

The *Pristine* Survey: CFHT ESPaDOnS Spectroscopic Analysis of 115 Bright Metal-Poor Candidate Stars

Originally published in Monthly Notices of the Royal Astronomical Society, Volume 492, Issue 3, p.3241-3262 (2020). Authors: Kim Venn, Collin Kielty, Federico Sestito, Else Starkenburg, Nicolas Martin, David Aguado, Anke Arentsen, Piercarlo Bonifacio, Elisabetta Caffau, Vanessa Hill, Pascale Jablonka, Carmela Lardo, Lyudmilla Mashonkina, Julio Navarro, Chris Sneden, Guillaume Thomas, Kris Youakim, Jonay I. González-Hernández, Rubén Sánchez Janssen, Ray Carlberg, Khyati Malhan

Personal Contributions

The spectra used in Venn et al. (2020) were delivered by CFHT, fully reduced using the Libre-Esprit pipeline. However, several data further reduction steps were required before these spectra could be used for a chemical abundance analysis. I wrote a python-based data reduction suite, which processes the individual visit spectra into a single, high SNR spectrum for each star. Starting from the individually extracted and normalized echelle orders, one continuous spectrum was stitched together by weighting the overlapping wavelength regions by their error spectrum, and coadding them via a weighted average. All visits for a given star were then coadded via a weighted mean using the error spectrum to provide the weights. The coadded spectrum was RV corrected through cross-correlation with a high SNR comparison spectrum of metal-poor star HD 122563, and the RV corrected spectrum is finally re-normalized using an asymmetric sigma-clipping routine. Further details are given in Section 2.3.

Gaia DR2 proper motions and astrometry have dramatically accelerated the fields of Galactic Archaeology and near-field cosmology by providing the data needed to calculate the detailed orbits of nearby stars. Enabled by these recent developments, my primary contributions to Venn et al. (2020) were in calculating kinematics and orbits for the 70 highly probable metal-poor candidates. I calculated orbital parameters for the sample with *Galpy* (Bovy, 2015), using Bayesian inferred distances (Sestito et al., 2019), the ESPaDOnS radial velocities, and *Gaia* DR2 proper motions as input data. The *MWPotential14*[1] was adopted, though a more massive halo of $1.2x10^{12}M_{\odot}$ was chosen following Sestito et al. (2019). Errors were propagated from the uncertainties in proper motion, RVs, and distance via Monte-Carlo sampling of the Gaussian distributions of the input quantities. These results are presented in Sections 2.6.1 - 2.6.3.

Additional contributions include the creation of Figures 2.2, 2.5, 2.7, 2.8, 2.9, 2.10, 2.16, 2.17, 2.18, 2.19, 2.20, 2.21, and A.1.

---

[1] This potential is three component model composed of a power law, exponentially cut-off bulge, Miyamoto Nagai Potential disc, and Navarro, Frenk & White (1997) dark matter halo.

2.1 Introduction

Very old stars are witness to the earliest epochs of galaxy formation and evolution. Most theoretical models of star formation at early times predict the formation of high mass stars (e.g., Nakamura & Umemura, 2001; Abel et al., 2002; Bromm, 2013) that contributed to the reionization of the Universe. During their short lives, these massive stars initiate the formation of the chemical elements beyond hydrogen, helium, and lithium, and yet no star with such a primordial composition has yet been found. Fragmentation of the early star forming regions has also been predicted (e.g., Schneider et al., 2003; Clark et al., 2011b; Greif, 2015; Hirano et al., 2015), providing an environment where lower mass ($\sim$ 1 $M_{\odot}$) stars could form, which would have much longer lifetimes. These old stars are expected to be metal-poor, having formed from nearly pristine gas, and could be used to trace the chemical elements from the massive (first) stars and their subsequent supernovae (e.g., Frebel & Norris, 2015; Salvadori et al., 2019; Hartwig et al., 2018).

In recent years, abundance patterns of metal-poor stars have been examined extensively (e.g., Keller et al., 2014; Nordlander et al., 2019; Ishigaki et al., 2018), pointing to the significance of low-energy (faint) supernovae, whose ejecta falls back onto their iron-cores, thereby mainly expelling light elements. It is not clear if these low-energy supernovae were more common at ancient times, or if concurrent massive stars underwent direct collapse to black holes and ceased nearby star formation, erasing any direct evidence of their presence in the next generation of stars. Overall, metal-poor stars allow us to examine nucleosynthetic yields from one or a few supernovae events to constrain the detailed physics of these events, such as neutron star masses, rotation rates, mixing efficiencies, explosion energies, etc.(Heger & Woosley, 2010; Thielemann et al., 2018; Wanajo, 2018; Müller et al., 2019; Jones et al., 2019). These yields are relevant for understanding the early chemical build-up and the initial conditions in the early Galaxy.

Chemical abundances also show variations between old metal-poor stars in different environments such as dwarf galaxies, suggesting that the first stages of enrichment were not uniform. Stars in the nearby dwarf galaxies typically have lower abundances of α- and odd-Z elements, attributed to their slower star formation histories and/or fewer number of high mass stars overall (Venn et al., 2004; Tolstoy et al., 2009; Nissen & Schuster, 2010; McWilliam et al., 2013; Frebel & Norris, 2015; Hayes et al., 2018), while significant variations in heavy r-process elements in some dwarf galaxies, and globular clusters, are discussed in terms of contributions from individual compact binary merger events, like GW170817 (e.g., Roederer, 2011; Roederer et al., 2018a; Ji et al., 2016b, 2019). In addition, about a third of the [Fe/H] < -2.5 stars[2] in the Galactic halo show very high enhancements in carbon (the carbon-enhanced metal-poor stars, "CEMP"; Yong et al. 2013; Aguado et al. 2019 also see Kielty et al. 2017; Mardini et al. 2019), discussed as a signature of the earliest chemical enrichment in the Universe. However, at least one ultra metal-poor star is not carbon-enhanced (SDSS J102915+172927, Caffau et al., 2012a), and the known metal-poor stars in the Galactic bulge do not show carbon-enhancements (Howes et al., 2016; Lamb et al., 2017). Norris et al. (2013) suggest that there are likely multiple chemical enrichment pathways for old metal-poor stars dependent on the star formation environment, and also possibly binary mass transfer effects (also see discussions by Starkenburg et al. 2014; Arentsen et al. 2019).

The majority of old, metal-poor stars in the Galatic halo are thought to have been accreted from dwarf galaxies at early epochs, based on cosmological hydrodynamical simulations of the Local Group (Ibata et al., 1994; Helmi et al., 1999; Ibata et al., 2004; Abadi et al., 2010; Starkenburg et al., 2017b; El-Badry et al., 2018a). This is consistent with the high-velocity, eccentric, orbits determined from the exquisite Gaia DR2 data (Gaia Collaboration et al., 2018) and spectroscopic radial velocities for a majority of the ultra metal-poor stars (Sestito et al., 2019) and the ultra faint dwarf galaxies (Simon, 2018). Interestingly, many of these orbits are also highly retrograde, similar to the diffuse halo merger remnants, Gaia-Enceladus (Helmi et al., 2018; Haywood et al., 2018; Belokurov et al., 2018; Myeong et al., 2018) and Gaia-Sequoia (Myeong et al., 2019; Barbá et al., 2019). However, some metal-poor stars have been found to have orbits that place them in the Galactic plane (Sestito et al., 2019), even with nearly circular orbits (e.g., SDSS J102915+172927, Caffau et al., 2012a). These latter observations challenge the cosmological simulations since metal-poor stars are assumed to be old, and yet the Galactic plane is thought to have formed only ~10 Gyr ago (e.g., Casagrande et al., 2016a; Gianninas et al., 2015). Alternatively, Sestito et al. (2019) suggest these stars may have be brought into the Galaxy from a merger that helped to form the disk.

Progress in this field will require large statistical samples of metal-poor stars in a variety of environments within the Local Group. Unfortunately, metal-poor stars are exceedingly rare and difficult to find, being overwhelmed by the more numerous metal-rich populations in the Galaxy. Examination of the Besançon model of the Galaxy (Robin et al., 2003),

---

[2]We adopt standard notation, such that [X/H] = $\log$(X/H)$_*$ − $\log$(X/H)$_\odot$.

which is guided by a theoretical framework for the formation and evolution of the main stellar populations, suggests that a typical halo field has only one in ~2000 stars with [Fe/H] < −3 between 14 < V < 18 (Youakim et al., 2017). Enormous effort has gone into the discovery and study of extremely, ultra, and hyper metal-poor stars with [Fe/H] < −3.0, [Fe/H] < −4.0, and [Fe/H] < −5.0, respectively. Most of the known metal-poor stars have been found in dedicated surveys, such as objective prism surveys (the HK survey and Hamburg-ESO survey, Beers et al., 1992; Beers & Christlieb, 2005; Christlieb et al., 2002a, 2008; Frebel et al., 2006; Schörck et al., 2009), wide-band photometric surveys (Schlaufman & Casey, 2014), and blind spectroscopic surveys, such as the the the Sloan Digital Sky Survey (SDSS) SEGUE and BOSS surveys (Yanny et al., 2009; Eisenstein et al., 2011; Dawson et al., 2013), and from the Large Sky Area Multi-Object Fibre Spectoscopic Telescope (LAMOST, Cui et al., 2012). According to the SAGA database (see Suda et al., 2017, and references therein), there are ~500 stars with [Fe/H] < −3.0, though fewer than half have detailed chemical abundances. Recently, narrow-band photometric surveys have shown higher success rates for finding metal-poor stars, particularly SkyMapper (Keller et al., 2007; DaCosta et al., 2019) and the *Pristine* survey (Starkenburg et al., 2017a; Youakim et al., 2017; Aguado et al., 2019). *Pristine* photometry with follow-up Keck II/DEIMOS spectroscopy has also been used to increase sample sizes and improve the chemodynamical studies of faint satellites (Draco II and Sgr II, Longeard et al., 2018, 2020). At the same time, Simon (2018) has shown that Gaia DR2 proper motion cleaning may also be a promising way to find new metal-poor members of ultra faint dwarf galaxies.

The *Pristine* survey uses a unique narrow-band filter centered on the Ca ɪɪ H & K spectral lines ("CaHK") mounted on MegaPrime/MegaCam at the 3.6-metre Canada France Hawaii Telescope (CFHT). When combined with broad-band SDSS *gri* photometry (York et al., 2000), this CaHK filter has been calibrated to find metal-poor candidates with 4200 < T < 6500 K. The *Pristine* survey has proven successful at predicting metallicities for faint objects (18 > V > 15), based on results from medium resolution spectroscopic follow-up (Youakim et al., 2017; Aguado et al., 2019). For brighter objects, the success of the *Pristine* calibration is less certain. Caffau et al. (2017) observed 26 bright (g < 15) candidates with the FEROS spectrograph at the MPG/ESO 2.2-metre telescope, but found only 5 stars with [Fe/H] < −2.0. It was thought that the selection may be affected by previously unrecognized saturation effects in the SDSS photometry. Thus, Bonifacio et al. (2019a) selected bright candidates using a new *Pristine* calibration with APASS photometry (c.f., APASS DR10 Henden, 2019); observations of 40 targets with the SOPHIE spectrograph at Observatoire de Haute Provence found only 8 stars with [Fe/H] < −2.0, and none with [Fe/H] < −3.0.

Until now, confirmation of the *Pristine* metallicity predictions below [Fe/H] = −3.0 has only been carried out for one star from high resolution spectroscopy, Pristine_221.8781+09.7844 at [Fe/H] = −4.7 (1D, LTE) and V = 16.4 (Starkenburg et al., 2018).

In this chapter, I present the analysis of 115 bright (V < 15) metal-poor candidates from the *Pristine* survey, calibrated using the original SDSS *gri* photometry and observed at the CFHT with the high resolution ESPaDOnS spectrograph. Such high resolution spectra are necessary for detailed chemical abundances, as well as precision radial velocities for determining the kinematic properties. The power of combining chemical abundances with kinematic properties of stars is the backbone of the field of Galactic Archaeology (e.g., Freeman & Bland-Hawthorn, 2002; Venn et al., 2004; Tolstoy et al., 2009; Frebel & Norris, 2015). We confirm the success of the *Pristine* survey to find metal-poor stars even at bright magnitudes, determine the chemical abundances for 10 elements, calculate the kinematics of the stars in our sample, and interpret in the context of variations in nucleosynthetic sites, locations, and time-scales. The study of metal-poor old stars is unique to our Local Group, since only here can we resolve individual stars and study these rare targets that guide our understanding of the physics of star formation, supernovae, the early build-up of galaxies, and the epoch of reionization.

2.2   Target Selection

Targets were selected from the *Pristine* survey catalogue[3], which includes 28557 bright (V < 15) stars in the original ~1000 sq.deg.^{2} footprint between 180 < RA < 256° and +00 < Dec < +16° (Starkenburg et al., 2017a; Youakim et al., 2017).

Pristine survey targets were cross-matched with SDSS photometry to obtain *ugri* broad-band magnitudes used for colour temperature determinations and point source identification. Additional selection criteria were adopted, as described by Youakim et al. (2017), including the removal of non-star contaminants (based on SDSS and CaHK flags), white dwarf contaminants (removing SDSS $u - g$ > 0.6, Lokhorst et al. 2016), variability flags from Pan-STARRS1 photometry (Hernitschek et al., 2016), and the quality of SDSS *gri*-band photometry. The SDSS *gri*-band photometry was further used for a colour selection, where $0.25 < (g-i)_o < 1.5$ and $0.15 < (g-r)_o < 1.2$ correspond to the temperature range 4200 K $< T_{\text{eff}} < 6500$ K, covering the tip of the red giant branch and the cooler main sequence to the main sequence turnoff.

---

[3]Internal-Catalogue-1802.dat

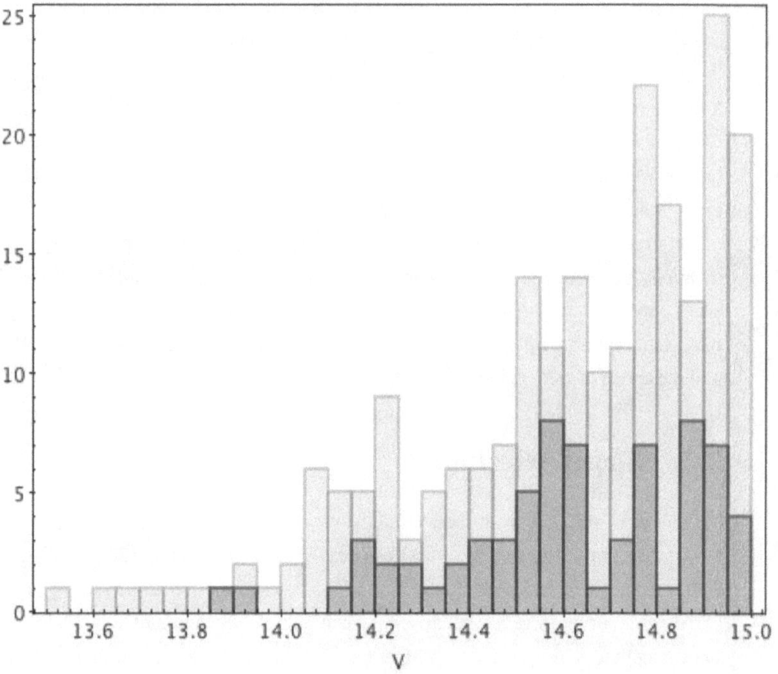

Figure 2.1 Histogram of the V magnitudes of 223 stars with high probabilities to be metal-poor ([Fe/H]< −2.50) from the (g-i) or (g-r) calibrations in the *Pristine* survey original ~1000 sq.deg.^{2} footprint (grey bars). The 70 stars observed with CFHT ESPaDOnS that also meet these criteria are overplotted (blue bars).

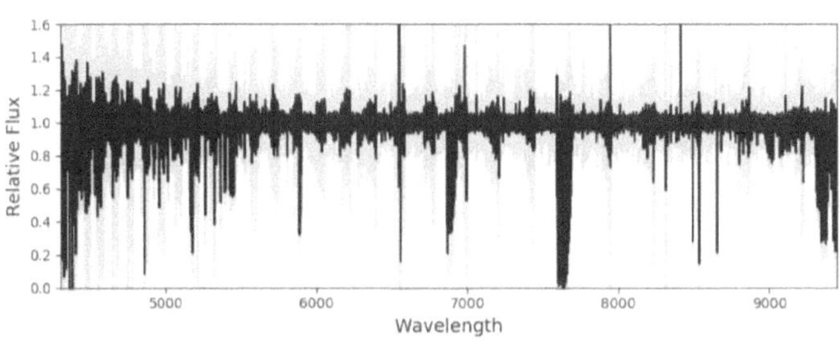

Figure 2.2 Full CFHT ESPaDOnS spectrum for Pristine-235.1449+08.7464 (grey), and smoothed by a 3-pixel boxcar (black). A signal-to-noise ratio (SNR) =30 near 520 nm was adopted for this exploratory survey, leaving very low SNR and non-smooth continuum on the red side of the detector and therefore in the inter-order regions.

The 115 stars observed at CFHT with the high resolution (R~68,000) ESPaDOnS spectrograph (Donati et al., 2006) are listed in Table D.1 including RA and DEC (from SDSS, in degrees), the dereddened SDSS $(ugri)_0$ and *Pristine*-CaHK$_0$ magnitudes, the V and I magnitudes (converted from SDSS photometry using Jordi et al. (2006) and not dereddened, thus in observer units), and the reddening E(B-V) value. Extinction values are small for most stars, and we assume that all the extinction is in the foreground, therefore using the Schlegel et al. (1998) extinction maps. A summary of the CFHT ESPaDOnS observing runs that comprise this program are 16AC031 (23 targets), 16AC096 (17 targets), 16BC008 (25 targets), 17AC002 (30 targets), 18BC018 (25 targets), which is 120 targets, with 5 repeat targets, thus 115 independent objects.

We note that this program began immediately after the initial *Pristine* survey MegaCam observing runs, and the metallicity calibrations have improved over the course of these spectroscopic follow-up observations (2016A to 2018B). Of the 115 observed stars, 88 remain in the *Pristine* survey catalogue. In Table D.1, we have 59 stars with >80% probability to have [Fe/H]< −2.25 using both the SDSS $g - r$ and $g - i$ colour calibrations, and with individual metallicity estimates of [Fe/H] < −2.5. Another 10 stars follow these selections using the SDSS $g - r$ colour alone. Youakim et al. (2017) showed that the SDSS i filter has saturation effects in some fields for stars in our magnitude range that can affect the SDSS $g - i$ selection criterion. An additional 46 stars were observed with ESPaDOnS, however we now recognize 19 of those to have low probabilities to be metal-poor, and 27 are no longer in the *Pristine* survey catalogue (e.g., due to the saturation effects in the SDSS photometry recognized later). Ironically, of those latter 27 stars, one star (Pristine_213.7879+08.4232) does appear to be metal-poor, e.g. its Ca ii triplet lines are weak and narrow. Possibly the *Pristine* survey selection function is now slightly overly strict; we retained this one metal-poor candidate. Thus we have observed a total of 70 (59 + 10 + 1) metal-poor candidates selected from the original ~1000 sq.deg.^{2} footprint of the *Pristine* survey. In total, there are 223 bright stars that meet all of the selection criteria described in this section, thus we have observed 31% (70/223) of these candidates. Both of these distributions are shown in Fig. 2.1.

The selection criteria used here differ slightly from Youakim et al. (2017) and Aguado et al. (2019), where stars with probability over 25% in both $g - r$ and $g - i$ were selected for their medium resolution spectroscopic program. These lower limits were also adopted by Caffau et al. (2017) and Bonifacio et al. (2019a) in their target selections, though using APASS photometry in the latter paper. We emphasize that our target selections were made without a priori knowledge of the spectroscopic metallicities, other than for a small subset of five stars[4] in our final 2018B observing run.

2.3 ESPaDOnS Observations

The CFHT high resolution spectrograph ESPaDOnS was used between 2016A and 2018B to observe 115 new bright, metal-poor candidates found in the original CFHT-MegaCam survey footprint as part of the *Pristine* survey. ESPaDOnS was used in the "star+sky" mode, providing a high resolution (R= 68, 000) spectrum from 400 to 1000 nm, making it possible to determine precision radial velocities, stellar parameters, and chemical abundances.

Each observation was fully reduced using the Libre-Esprit pipeline[5]. This included subtraction of a bias and dark frames, flat fielding for pixel to pixel variations, and masking bad pixels. ESPaDOnS records 40 orders, each one of them curved, such that Libre-ESpRIT performs a geometric analysis from the calibration exposures before it performs an optimal extraction. It also corrects for the tilt of the slit, determines the wavelength calibration from a thorium lamp exposure, and applies the heliocentric correction. The "star+sky" mode enables good sky subtraction during the pipeline reductions. The final (combined) spectra were renormalized using an asymmetric k-sigma clipping routine.

As this is an exploratory program, spectra were collected until signal-to-noise SNR>30 near 520 nm was reached per target; multiple exposures were coadded for fainter targets to reach this SNR. A full sample spectrum for one metal-poor target is shown in Fig. 2.2, where it can be seen the SNR worsens at shorter wavelengths. In addition, the red side of the CCD detector in this cross-dispersed echelle spectrograph is less illuminated than the centre of each order, causing lower SNR in the interorder regions. Overall, this impacts the smoothness of the spectra. Spectral lines in the low SNR regions were rejected from this analysis. In total, this observing campaign used over 150 hours of CFHT ESPaDOnS time.

Radial velocities (see Tab. D.2) were determined by fitting several strong lines per star, and averaging the results from the individual lines together. This method was selected rather than a more rigorous use of a cross correlation technique (e.g., IRAF/*fxcorr*) because of slight wavelength solution variations for lines in common between orders and the significant noise in the interorder regions. The typical uncertainty in RV is σRV ≤ 0.5 km s^{-1} for

---

[4]Five stars had interesting results from our concurrent medium resolution spectral campaign, and were selected for observations with ESPaDOnS during our final 2018B run. Three were confirmed to be metal poor ([Fe/H]< -2.5), but two were not ([Fe/H]> -2.0). If we recalculate our success rates without these five stars, then 38% (25/65) are found with [Fe/H]< -2.5 and 16% (4/25) with [Fe/H]< -3.0.

[5]Libre-ESpRIT is a self-contained data reduction package developed specifically for reducing the ESPaDOnS echelle spectropolarimetric data developed by Donati et al. (1997)

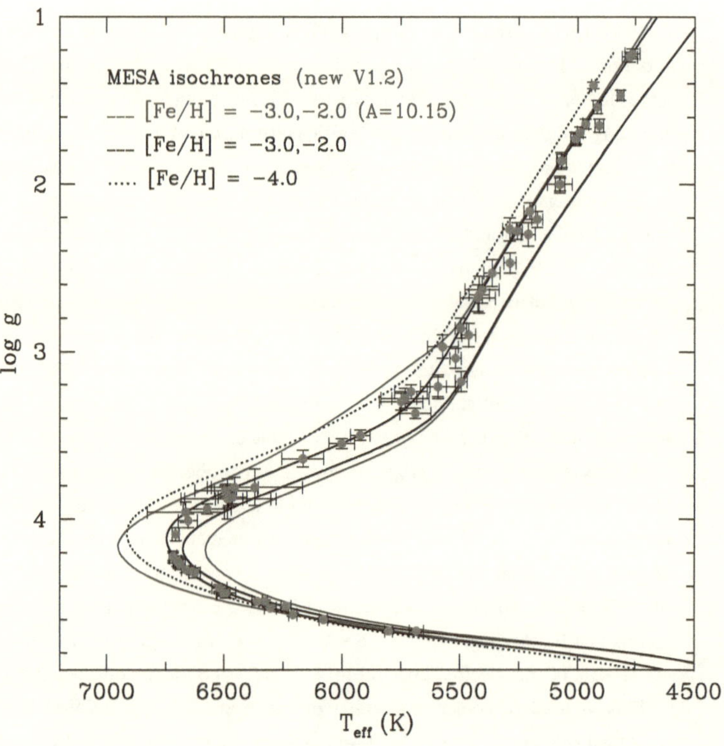

Figure 2.3 T_{eff} vs log g for 70 high-probability metal-poor stars selected from the *Pristine* survey. For illustration purposes, the isochrones for a single age of 14.1 Gyr are shown (or log(A/yr) = 10.15). The isochrones used for the stellar parameter estimates are from a previous version of MESA/MIST (shown in black), compared with isochrones from the newer version of MIST (V1.2, shown in blue).

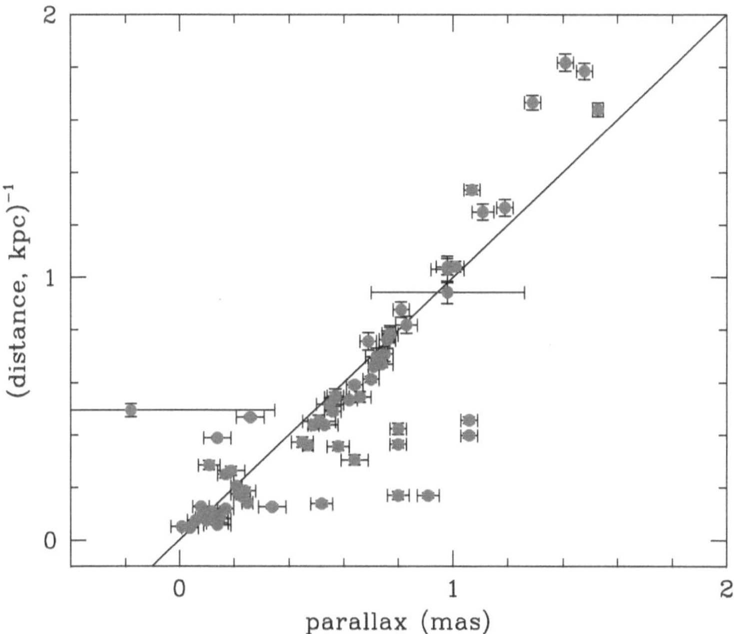

Figure 2.4 A comparison of the Gaia DR2 parallax measurements (with zero point correction, see text) and 1/(distance, in kpc) from the Bayesian inference method developed by Sestito et al. (2019) for our 70 metal-poor candidates.

lines below 6000 Å. Variations between the RV solutions were noticed between the CaT lines (~8500 Å) vs lines in the blue (below 6000 Å), ranging from 0 to 3 km s^{-1}. A similar offset was seen in CFHT ESPaDOnS spectra for CEMP stars by Arentsen et al. (2019), who showed that the RV derived from lines below 6000 Å provide better agreement with radial velocity standards. Therefore, we did not use any lines above 6000 Å for the RV measurements. The variations for common lines in overlapping orders was small (1-2 pixels, or ≤ 0.8 Å per line); when averaged over several lines (>10) this intrinsic variation corresponds to ≤ 0.5 km s^{-1}, the RV uncertainty that we adopt for all of our spectra. Multiple observations were spaced over a narrow range in time, so that no RV variability information is available for identifying potential binary systems.

2.4 Spectroscopic Analysis

The analysis of stellar spectra requires a comparison with synthetic spectral calculations of the radiative transfer through a model atmosphere. In this study, we adopt the ATLAS12 (Kurucz, 2005) and MARCS (Gustafsson et al., 2008) 1-D, hydrostatic, plane parallel models, in local thermodynamic equilibrium. These models are represented by an effective temperature (T_{eff}), surface gravity (log g), mean metallicity (represented as the iron abundance, [Fe/H]). The model atmospheres are generated with scaled solar abundances, but increased α element abundances to represent the majority of metal-poor stars in the Galaxy ([α/Fe]=0.0 to +0.4). Microturbulence (ξ) is assumed to scale with gravity, using the scaling relations by Sitnova et al. (2015) and Mashonkina et al. (2017a) for Galactic metal-poor dwarfs and giants, respectively.

Initial stellar parameters (temperature and metallicity) were determined from photometry. A colour temperature was determined from the SDSS *gri* colours and the semi-empirical calibrations from González Hernández & Bonifacio (2009), and metallicity was determined from the SDSS *gri* photometry with the *Pristine* CaH&K filter, with calibrations described by Starkenburg et al. (2017a). Our targets range in colour temperature (=T_{SDSS}) from 4700 to 6700 K, and have *Pristine* metallicities [Fe/H]$_{Pristine} \lesssim -2.5$; see Tab. D.1.

2.4.1 Stellar parameters using SDSS and Gaia DR2 data, and MIST isochrones ("Bayesian inference" method)

Improved stellar temperatures and the gravity estimates were determined using a "Bayesian inference" method developed by Sestito et al. (2019). A probability distribution function

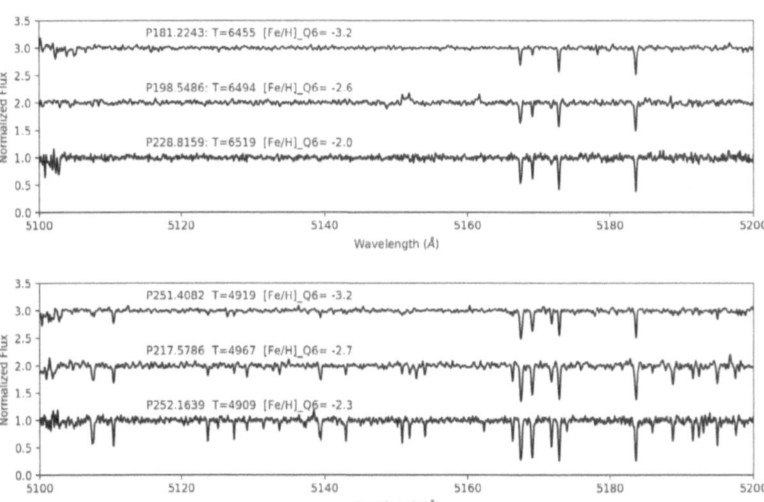

Figure 2.5 Sample CFHT ESPaDOnS spectra for three hot (T~6500 K) main sequence turn-off stars and three cool (T~4900 K) red giants. Each spectrum is labelled with the target name, temperature from the Bayesian inference method, and spectroscopic metallicity from our "Quick Six" analysis.

of the heliocentric distance to each star was inferred by combining the SDSS photometric colours and Gaia DR2 parallaxes data, with stellar isochrones, and a Milky Way stellar density prior. We apply the zero point offset on the parallax of -0.029 mas recommended by Lindegren et al. (2018), but note that the Gaia team have discussed the possbilitiy of spatially correlated parallax errors ranging from 0.1 to 0.01 mas; see discussion by Zinn et al. 2019. Isochrones are from the MESA/MIST library (Paxton et al., 2011; Dotter, 2016; Choi et al., 2016), which reach the lowest metallicities ([Fe/H] ≤ -4); see Fig 2.3. A flat age prior was assumed between 11 and 14 Gyr (or log(A/yr)= 10.05 to 10.15), and we adopted a gaussian PDF for the metallicity centred on the *Pristine* photometric metallicity.

Unique solutions for the stellar parameters were found for 85 of our targets (out of 89 stars; the 88 stars that remain in the *Pristine* survey catalogue after photometric quality cuts, and one star that we have retained, see Section 2.2). Another four stars have sufficiently large parallax errors that we could not distinguish between the dwarf or giant solutions; both are given in Tab. D.2. It is recognized that determining the distance to a star simply by inverting the parallax measurement can lead to substantial errors, especially when the parallax is small (or even negative), and when there is a relatively large measurement uncertainties (e.g., $\Delta\pi/\pi > 0.1$); see Bailer-Jones et al. (2018). One advantage of this Bayesian inference method is that stars with negative parallax results and stars with very large parallax errors can be placed onto the isochrones and assumed to be distant. In Fig. 2.4, the Bayesian inferred distances are compared to the Gaia DR2 parallax measurements.

For two stars (Pristine_200.5298+08.9768 and Pristine_187.9785+08.7294), the Bayesian inferred distance method seemed to fail, placing these stars in the outer Galactic halo, even though they have relatively large parallax measurements with small uncertainties (0.46 ± 0.04 mas and 0.74 ± 0.04 mas in the Gaia DR2 catalogue), and they are metal-rich (e.g., visibly strong Ca II triplet lines). Since we had assumed these stars are metal-poor (from their *Pristine* metallicities), then the metal-poor isochrone used to compute their distances was incorrect, and resulted in a poor distance estimate. By adjusting their distances to simply 1/parallax (i.e., not using the metal-poor stellar isochrones), then both of these stars are located closer to the Sun, consistent with the majority of metal-rich stars in the Galaxy. For our main targets, stars that the *Pristine* survey identifies as metal-poor and that are truly metal-poor, then this will not be a problem, and we expect this Bayesian inference method will provide

2.4.2 Initial ("Quick Six") spectroscopic metallicities

Adopting the stellar parameters from the Bayesian inference method described above (Section 2.4.1), then a model atmosphere was generated from both the MARCS and ATLAS grids. Elemental abundances were computed using a recent version of the 1D LTE spectrum analysis code MOOG (Sneden, 1973; Sobeck et al., 2011).

As an initial spectroscopic metallicity estimate, a subset of six iron lines were selected that are observable in the good SNR regions of the ESPaDOnS spectra; 4x Fe $\textsc{i}$ (λ4957.6, λ5269.5, λ5371.5, λ5397.1) and 2x Fe $\textsc{ii}$ (λ4923.9, λ5018.4). These are well-known and fairly isolated spectral lines, with good atomic data[6] and line strengths across the parameter range. The equivalent widths of these six lines were measured using IRAF/*splot*[7], measuring both the area under the continuum and by fitting a Gaussian profile, comparing the results. We call the average of these six LTE line abundances our "Quick Six" spectroscopic metallicities ([Fe/H]$_{Q6}$), and these are used as an initial test of the *Pristine* metallicity estimates.

Sample spectra are shown for six targets; three hot (T$\sim$6500 K), main sequence turn-off stars and three cool (T$\sim$4900 K) red giants in Fig. 2.5. These spectra are labelled with their target name, temperature (from the Bayesian inference method, see Section 2.4.1), and metallicity [Fe/H]$_{Q6}$ from this "Quick Six" analysis.

Departures from LTE are known to overionize the Fe $\textsc{i}$ atoms due to the impact of the stellar radiation field, particularly in hotter stars and metal-poor giants. These non-LTE (NLTE) effects can be significant in our stellar parameter range, such that NLTE corrections typically reduce the line scatter and improve the Fe $\textsc{i}$=Fe $\textsc{ii}$ ionization balance (Amarsi et al., 2016; Sitnova et al., 2015; Mashonkina et al., 2019). NLTE effects are explored in this "Quick Six" analysis, by comparing the results from Mashonkina et al. (2017a, 2019) and the INSPECT table[8] (Amarsi et al., 2016; Lind et al., 2012). INSPECT provides data for one of the selected lines, Fe $\textsc{i}$ λ5269, where the NLTE correction is Δ(Fe $\textsc{i}$) $\leq$ 0.15, over our parameter space, where Fe $\textsc{i}$(NLTE) = Fe $\textsc{i}$(LTE) + Δ(Fe $\textsc{i}$). Based on a similar treatment

---

[6]Atomic data for the Fe $\textsc{i}$ lines are from Blackwell et al. (1979a) with high precision, or from the laboratory measurements from O'Brian et al. (1991). The Fe $\textsc{ii}$ lines have less certain atomic data from Raassen & Uylings (1998), however a NLTE investigation by Sitnova et al. (2015) showed that these lines have tiny NLTE corrections *and* yield iron abundances in metal-poor stars within 0.1 dex of all other Fe $\textsc{i}$ and Fe $\textsc{ii}$ lines that they studied. We also note Roederer et al. (2018b) used 3 Fe $\textsc{i}$ and 1 Fe $\textsc{ii}$ of these lines in their detailed iron analysis of six warm metal-poor stars.

[7]IRAF is distributed by the National Optical Astronomy Observatories, which is operated by the Association of Universities for Research in Astronomy, Inc. (AURA) under cooperative agreement with the National Science Foundation

for inelastic collisions (of Fe I with H I), Mashonkina et al. (2017a) predict similar NLTE corrections for the other three Fe I lines ($\lambda 4957$, $\lambda 5372$ and $\lambda 5397$). The largest NLTE corrections (Δ(Fe I) $\sim$ 0.3) are for stars on the subgiant branch, while main sequence stars have $\sim$zero corrections. Recently, Mashonkina et al. (2019) have examined the impact of quantum mechanical rate coefficients on the inelastic collisions, and find that the NLTE corrections could be even larger (more positive) in the atmospheres of warm metal-poor stars, but smaller (even negative) in cool metal-poor stars and with a wide variation depending on the specific spectral line. This suggests that the NLTE calculations for Fe I need further study; however, given that these corrections in the literature are smaller than or equal to our measurement errors, then we do not apply the NLTE corrections in this "Quick Six" analysis.

The Fe I and Fe II individual line abundances are averaged together to find $[\text{Fe/H}]_{\text{Q6}}$ and the standard deviation $\sigma[\text{Fe/H}]_{\text{Q6}}$. Each of these results and the total number of lines used (≤ 6) are shown in Table D.2. From this analysis, we find that several of the *Pristine* metal-poor candidates are not metal-poor. A comparison of the $[\text{Fe/H}]_{\text{Q6}}$ iron abundances to the $[\text{Fe/H}]_{\text{Pristine}}$ predictions are shown in Fig. 2.6. These results are similar to the medium resolution spectral analyses (Youakim et al., 2017; Aguado et al., 2019), and discussed further in Section 2.4.5.

2.4.3 Comparing stellar temperatures

A comparison of stellar temperatures from the Bayesian inference method (Section 2.4.1) to the SDSS colour temperature (T_{SDSS}) is shown in Fig. 2.7. T_{SDSS} were the initial temperature estimates calculated using the InfraRed Flux Method[9], assuming [Fe/H] = -2.5, and based on SDSS (g-i) photometry. An average of the dwarf and giant solutions was used. For 10 stars, their ($g - i$) colours are unreliable because of saturation flags, and we adopt the relation based on the ($g - r$) colours from Ivezić et al. (2008). With this relation, a 200 K offset was applied to move from [Fe/H] = -0.5 to [Fe/H] = -2. Thus, we expect these values of T_{SDSS} to be an oversimplification, and are not surprised by the comparisons in Fig. 2.7, which are colour-coded by the "Quick Six" metallicities $[\text{Fe/H}]_{\text{Q6}}$.

Ignoring the metal-rich stars, then there is still a systematic offset between these methods for the metal-poor stars: the T_{SDSS} colour temperatures are too hot by $\sim$150 K for stars between T = 4700 $-$ 5700 K, but they are too cool by $\sim$200 K for stars with T > 6000 K. This offset is similar to the uncertainties in the Bayesian inference method temperatures

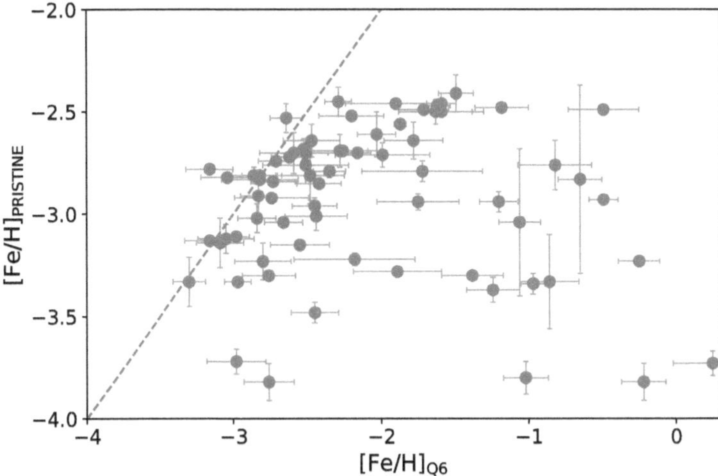

Figure 2.6 Comparisons of the "Quick Six" [Fe/H]$_{Q6}$ spectral abundances compared with the *Pristine* [Fe/H]$_{Pristine}$ photometric predictions. Clearly some of the *Pristine* metal-poor candidates are not metal-poor stars.

(T$_{Bayes}$) for most stars, where σT$_{Bayes}$ ranges from ~10 to 200 K (Table D.2). The very small colour temperature errors dT$_{SDSS}$ $\lesssim$10 K in Table D.1 are based on the difference between the dwarf/giant solutions, and are not realistic uncertainties.

2.4.4 Comparing gravity and Fe I=Fe II

Ionization balance has traditionally been used as an indicator of surface gravity in a classical model atmospheres analyses. Therefore, we compare the log g values from the Bayesian inference method (Section 2.4.1) to the difference in the [Fe I] and [Fe II] abundances, in Fig. 2.8. This figure is colour-coded by the "Quick Six" spectroscopic metallicities ([Fe/H]$_{Q6}$). For the metal-poor stars, the majority of our stars show Fe I=Fe II to within 2σ of the measurement errors, with a mean offset of [Fe I] − [Fe II] = +0.2. The measurement errors are calculated as the line weighted average of Fe I and Fe II.

For stars with poor agreement between iron ionization states, the cause cannot be due to neglected NLTE effects which appear to increase the Fe I abundance even further (see in Section 2.4.2). The offset is primarily seen in the cooler stars in our sample that are

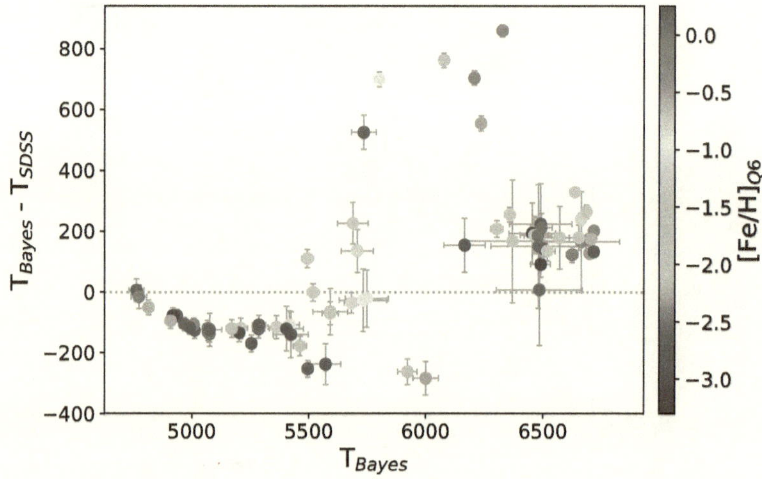

Figure 2.7 Comparisons of the *Pristine* survey colour temperature (T_{SDSS}) and the effective temperature determined from Bayesian inference method (T_{Bayes}) for our 70 metal-poor candidates. The data points are coloured by their metallicities from our spectroscopic [Fe/H]$_{Q6}$ analysis. As both temperature estimates adopte the *Pristine* photometric metallicity estimates [Fe/H]$_{Pristine}$ a priori, then clearly the metal-rich stars are not well calibrated.

on the red giant branch (with lower gravities). For these stars, the NLTE corrections are expected to be small (Δ(Fe I)$\lesssim$0.15). For stars closer to the main sequence turn off, the NLTE corrections can be larger; however, the offset between the Fe I and Fe II abundances seems smaller for those stars in our results. Therefore, the source of ionization equilibrium offsets is not yet clear.

For the metal-rich stars, we expect the surface gravities to be unreliable since the photometric *Pristine* metallicities [Fe/H]$_{Pristine}$ were assumed a priori in the Bayesian inference method. We do not investigate the metal-rich stars beyond our "Quick Six" analysis.

2.4.5 Comparisons with MRS analyses (FERRE)

A simultaneous *Pristine* survey program has been carried out for fainter stars ($15 < V < 17$) with medium resolution ($R \sim 1800$) spectroscopy at the 2.4-m Isaac Newton Telescope (INT), 4.2-m William Herschel Telescope, and 3.6-m New Technology Telescope (Aguado et al., 2019). These spectra have been observed with uniform spectral wavelength coverage, 360-550 nm, and analysed using the ASSET synthetic spectral grid (Koesterke et al., 2008). Both the observed and the synethetic spectra have been continuum normalized with the same functions, and the χ^2 minimization algorithm FERRE (Allende Prieto et al., 2006) is applied to derive the stellar parameters (temperature, gravity, metallicity, carbonacity).

The most recent analysis of the medium resolution spectroscopic data includes 946 stars (Aguado et al., 2019), where 13 of those stars are also in our sample of 70 high probability metal-poor stars (recall, that only 5 were observed at the INT first, and did not affect our target selections). In Figs. 2.8, 2.9, and 2.10, the surface gravities, temperatures, and metallicities are compared between the two analyses for the 13 stars in common. The large differences in gravity are from the *systematic* errors in the medium resolution FERRE analysis. While the FERRE analysis struggles with precision gravities, both methods are still able to break the dwarf-giant degeneracies sufficiently.

There is a clear relationship between the temperatures such that those determined from isochrones in the Bayesian inference method are cooler by $\sim$200 K near 5000 K and hotter by $\sim$500 K near 6700 K compared to the FERRE temperatures. These offsets are slightly smaller when compared with the SDSS colour temperatures T_{SDSS}. These temperature differences correlate with small-to-moderate metallicity offsets (Δ[Fe/H]$\leq$ 0.3) for stars cooler than 6000 K, whereas two of the hotter stars show larger metallicity offsets, Δ[Fe/H] $\sim$ 0.5. In summary, this analysis adopts the stellar parameters from the Bayesian inference method, and finds that the hot stars are hotter and less metal-poor than the results from the medium resolution FERRE analysis.

2.5 New stars with [Fe/H]$\leq$ -2.5

We have identified 28 new metal-poor stars, with spectroscopic metallicity [Fe/H]$_{Q6} \leq -2.5$, and where both [Fe I/H] and [Fe II/H] are below -2.5 dex (with the exception of Pristine_198.5486+11.4123, with [Fe I/H]$= -2.42$, which we retain because of its interesting orbit, discussed below). In this section, a more complete LTE, 1D model atmosphere analysis is carried out for a larger set of spectral lines and chemical elements.

As a comparison star, a spectrum of HD 122563 from the CFHT archive was analysed using the same methods as for the *Pristine* survey targets. Its metallicity is adopted from the literature, i.e., [Fe/H] $= -2.7 \pm 0.1$ (see Collet et al., 2018, and references therein), and our methods using its SDSS colours and Gaia DR2 parallax measurements yield stellar parameters that are in good agreement with the literature: $T_{eff}= 4625 \pm 50$ K, $\log g=$

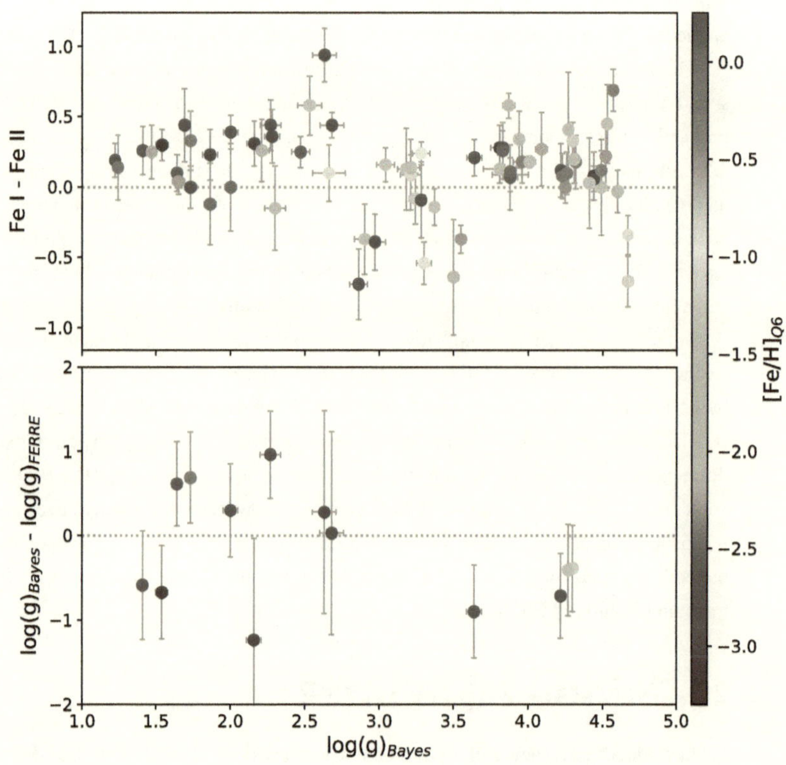

Figure 2.8 Comparisons of the surface gravities and iron ionization balance estimates for our 70 metal-poor candidates from the *Pristine* survey (top panel), and comparisons of our surface gravities versus those from the FERRE analysis of medium resolution spectra (Aguado et al., 2019) for 13 stars in common. The uncertainties in the gravities from FERRE can be quite large for the metal-poor stars due to a lack of suitable spectral signatures. The data points are coloured by their metallicities from our spectroscopic [Fe/H]$_{Q6}$ analysis.

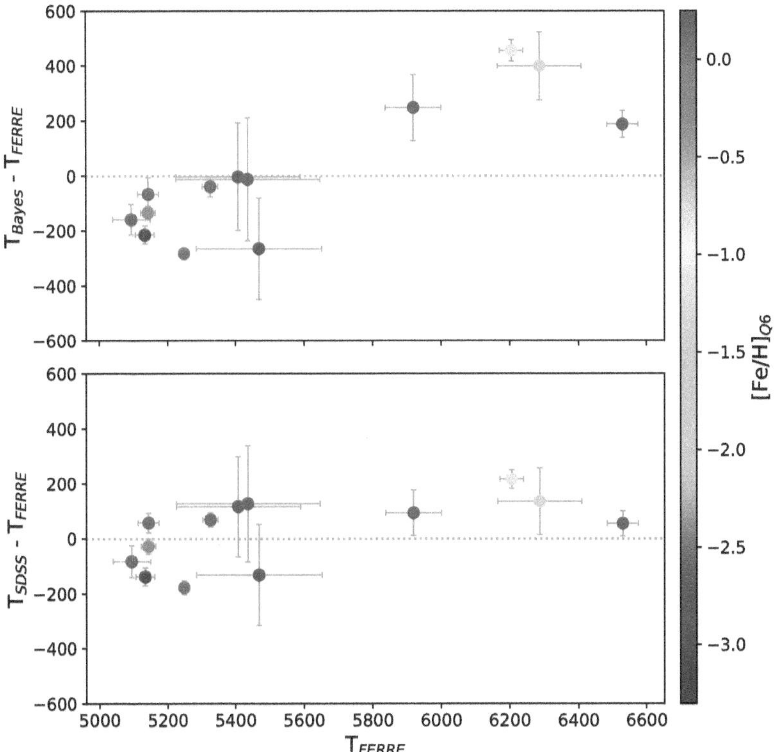

Figure 2.9 Temperature comparisons for 13 stars in common between the Bayesian inference analysis of our CFHT ESPaDOnS spectra and the FERRE analysis of medium resolution spectra (top panel, Aguado et al., 2019). The temperature offsets are slightly smaller when compared with the *Pristine* colour temperatures (T_{SDSS}, bottom panel).

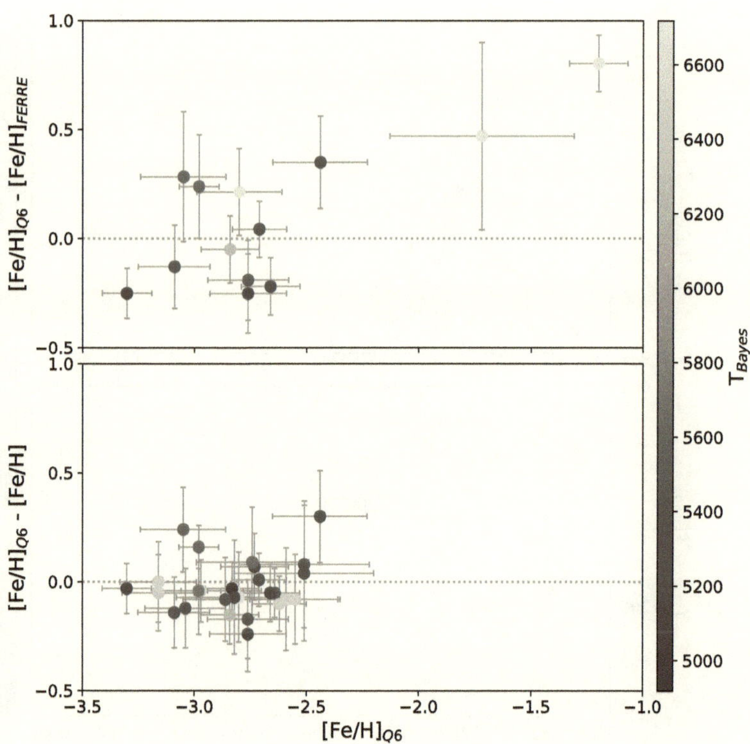

Figure 2.10 Metallicity comparisons for 13 stars in common between the [Fe/H]$_{Q6}$ analysis of our CFHT ESPaDOnS spectra and the FERRE analysis of medium resolution spectra (top panel, Aguado et al., 2019). [Fe/H]$_{Q6}$ values are also compared to the improved [Fe/H] values for our 28 very metal-poor stars, which include more lines of both Fe I and Fe II. The errors in the bottom panel are dominated by the "Quick Six" σ[Fe/H]$_{Q6}$. analysis.

1.6 ± 0.1. Microturbulence (ξ) was set to 2.0 km s^{-1} using the relationship with gravity from Mashonkina et al. (2017a).

For all 28 *Pristine* survey stars and HD122563, we identify and measure as many clean spectral lines as possible for a detailed abundance analysis, including more lines of Fe I and Fe II for higher precision iron abundances (than from the [Fe/H]$_{Q6}$ analysis). Starting with the spectral line list from Norris et al. (2017), spectral features were identified and measured using DAOSpec (Stetson & Pancino, 2008), and frequently checked by measuring the area under the continuum using IRAF/*splot*. Atomic data were updated when appropriate by comparing to the *linemake* atomic and molecular line database[1]. Abundance results from the model atmospheres analysis are compared to the solar (photospheric) abundances from Asplund et al. (2009).

2.5.1 Iron-group elements

The 28 new very metal-poor stars were initially identified from their [Fe/H]$_{Q6}$ abundances in Table D.2.

The iron abundances have been recalculated from 2-86 lines of Fe I, 2-6 lines of Fe II; see Table D.3. A 3σ minimum equivalent width was used to calculate an upper limit for Fe II for one star. The line-to-line scatter in the Fe I abundances range from σ(Fe I) = 0.12 to 0.24, even when only a small number of lines were measured. This is noteworthy because when other elements have < 4 lines, we adopt the larger of σ(X) or σ(Fe I)/$\sqrt{(N_X)}$ as a better representation of their line scatter.

These extended iron line measurements and abundances are not used to redetermine the spectroscopic stellar parameters for three reasons: (1) low sensitivity to the precise metallicity in the Bayesian inference method for the confirmed metal-poor stars, (2) insufficient number of lines of Fe II (and often Fe I) for a fully independent analysis, and (3) the SNR of our CFHT ESPaDOnS spectra ($\leq$ 30) is such that individual measurements of weak lines remain somewhat uncertain. The total iron abundance [Fe/H] is calculated as a weighted mean of Fe I and Fe II, and the total error δ[Fe/H] as σ([Fe/H])/(NFe I+NFe II)$^{1/2}$. These iron abundances are shown in Fig. 2.11 (top panel), where the errorbars include the systematic errors from the stellar parameter uncertainties added in quadrature (see Section 2.5.5, though the systematic errors tend to be much smaller).

---

[1]*linemake* contains laboratory atomic data (transition probabilities, hyperfine and isotopic substructures) published by the Wisconsin Atomic Physics and the Old Dominion Molecular Physics groups. These lists and accompanying line list assembly software have been developed by C. Sneden and are curated by V. Placco

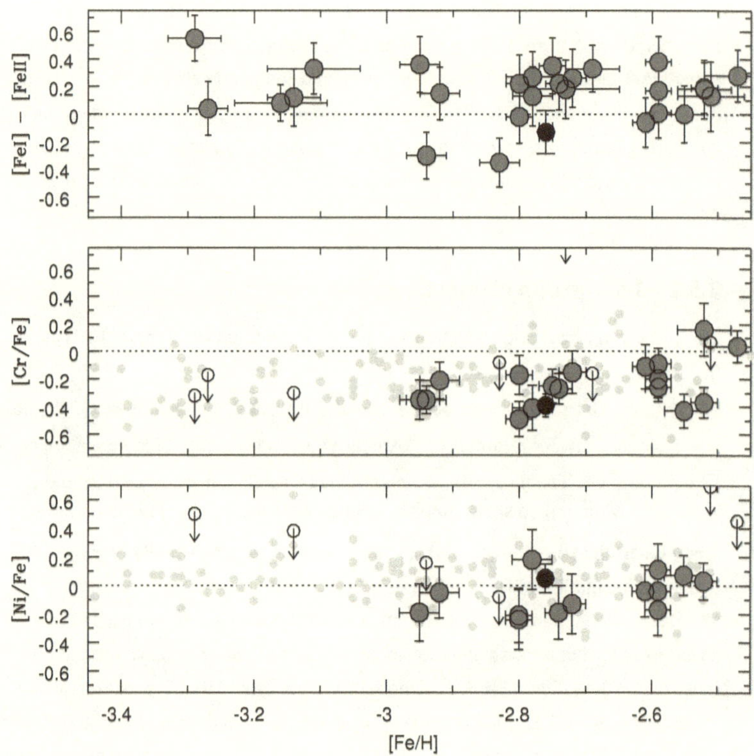

Figure 2.11 Iron-group (Fe, Cr, Ni) abundances and upper limits in our 28 new very metal-poor stars ([Fe/H] < −2.5, red points). Analysis results of the CFHT ESPaDOnS spectrum for the standard star HD 122563 are included (black point). Errorbars are the measurement errors and systematic errors combined in quadrature. Galactic comparison stars are included from the homogeneous analysis by Yong et al. (2013, small grey points).

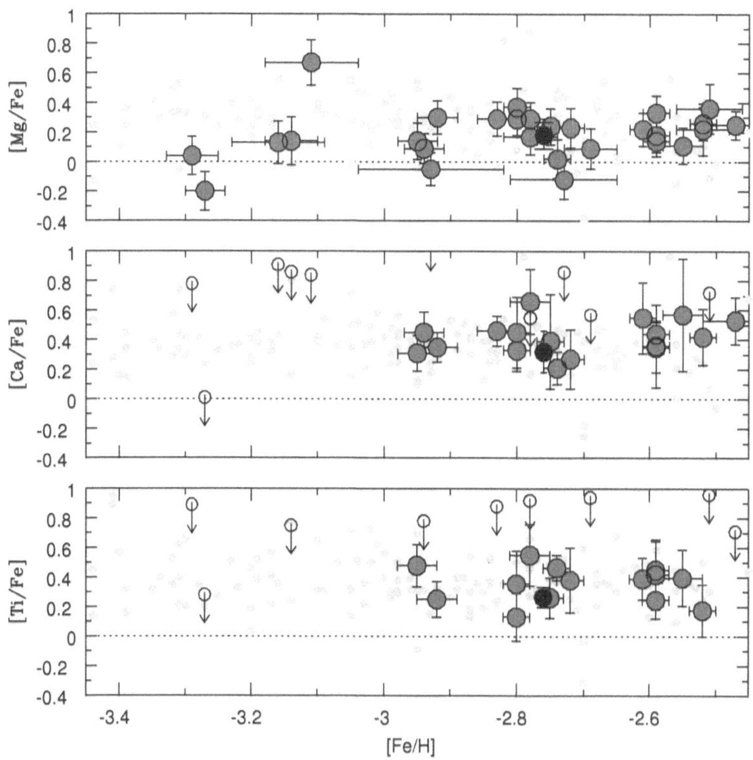

Figure 2.12 Mg, Ca, and Ti abundances and upper limits in the 28 new metal-poor stars ([Fe/H] < −2.5); symbols the same as in Fig. 2.11.

There is good to fair agreement between Fe I and Fe II, such that [Fe I] − [Fe II] ranges from $\sim$ −0.2 to +0.2. There is a median offset $\sim$ +0.2 for the sample, which is *not* due to NLTE corrections (see the discussion in Section 2.4.2). This may be due to the lack of Fe II lines in our metal-poor stars spectra for robust measurements, but another possibility is a systematic gravity uncertainty $\Delta log\ g \sim$ 0.5. High resolution spectra at bluer wavelengths (4000 Å) would provide more lines of Fe II to test this in the future. We also examine the slopes in the Fe I line abundances vs excitation potential (χ, in eV) to test our temperature estimates. A meaningful slope could be measured when N(Fe I)> 15 and $\Delta\chi$ > 3 eV, and all slopes were found to be relatively flat, < 0.1 dex/eV. This gives us more confidence in the fidelity of the temperatures T_{Bayes}, and thereby the Bayesian inference method for calculating stellar parameters and uncertainties.

The other iron-group elements (Cr and Ni, listed in Table D.3) are in good agreement with [Fe/H], and/or other Galactic halo stars at similar metallicities; see Fig. 2.11. Cr is determined from 1-3 lines of Cr I (5206.0, 5208.4 , and 5409.8 Å); only the spectrum of Pristine_245.8356+13.8777 had sufficient SNR at blue wavelengths that the lines at 4254.3, 4274.8, and 4289.7 Å could also be observed. [Cr/Fe] is subsolar in metal-poor stars, suggested as a NLTE effect (Bergemann & Cescutti, 2010). Ni is determined from 1-2 lines of Ni I (5035.4, 5476.9 Å). Three additional lines were available in the high SNR spectrum of HD 122563 (5080.5, 6643.6, and 6767.8 Å). The [Ni/Fe] results are within 1σ of the solar ratio, similar to other Galactic halo stars.

The α-element abundances (Mg and Ca) in the 28 new very metal-poor stars are listed in Table D.4. Upper limits are determined for some stars by computing 3σ minimum equivalent widths. The α-elements form through hydrostatic H- and He-core burning stages, though some Ca can also form later during SN Ia events. Because of these different nucleosynthetic sites, the [Mg/Ca] ratio need not scale together at all metallicities, as seen in some dwarf galaxies such as the Carina and Sextans dwarf galaxies (e.g., Norris et al., 2017; Jablonka et al., 2015; Venn et al., 2012), also the unusual star cluster NGC 2419 (Cohen & Kirby, 2012). We also include our discussion of Ti in this section even though it does not form with the α-elements. The dominant isotope ^{48}Ti forms primarily through Si-burning in massive stars (e.g., Woosley et al., 2002), and yet it seems to scale with other α-elements in metal-poor stars in the Galaxy.

Mg is determined from 2-3 lines (5172.7, 5183.6, 5528.4 Å), and a fourth line (4703.0 Å) was measurable in one star (Pristine_245.8356+13.8777). In Fig. 2.12, a larger scatter can be seen in the [Mg/Fe] results, though this is similar to the Galactic comparison stars. One star shows sub-solar [Mg/Fe] by more than 1σ (Pristine_251.4082+12.3657). Another star

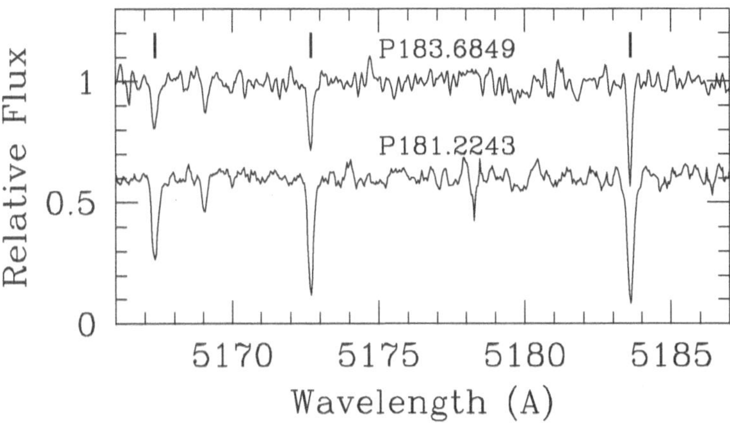

Figure 2.13 The spectrum of the Mg I b lines in the Mg strong star, P181.2243. This star is compared to Pristine_183.6849+04.8619, which has very similar stellar parameters (T~6450, log(g)~4, [Fe/H]~-3.2), but is Mg normal.

has high [Mg/Fe]~ +0.6, validated from all three Mg I lines (Pristine_181.2243+07.4160), also shown in Fig 2.13.

2.5.2 α-elements

The calcium abundances are determined from 1-9 lines of Ca I. The [Ca/Fe] abundances are in good agreement with each other, and with the Galactic comparison stars, as seen in Fig. 2.12. The same star with low [Mg/Fe] (Pristine_251.4082+12.3657) also has a very low [Ca/Fe] upper limit. This star is discussed further in Section 6.2.

Titanium has been determined from 1-9 lines of Ti I and 2-11 lines of Ti II. When both are unavailable, upper limits are determined from two Ti II lines (which provide stronger constraints than the Ti I features). In Fig. 2.12, the unweighted average results of [Ti I/Fe] and [Ti II/Fe] are shown.

NLTE corrections have *not* been incorporated for Mg, Ca, or Ti because they tend to be small to negligible ($\Delta \leq 0.1$ dex) according to the INSPECT database (for Mg I) and Mashonkina et al. (2017d, for Ca I). For Ti I, three lines (4981.7, 4991.1, 4999.5 Å) are available in the INSPECT database, which suggests large corrections $\Delta \sim +0.5$ dex. However NLTE corrections for the same lines from Sitnova et al. (2016), using a model atom that includes important high excitation levels of Ti I, are significantly smaller, $\Delta \sim +0.2$ dex. NLTE corrections should be included, but most of our stars have Ti I $\sim$ Ti II to within 1σ (our measurement errors) in LTE. Therefore, for this analysis, where the maximum SNR per star is ≤ 30, we do not include the small NLTE corrections, and note that the good agreement with the Galactic comparison stars and Ti ionization balance furthers our confidence in the stellar parameters from the Bayesian inference method.

No oxygen abundances or upper limits were determined since the [OI] 6300 and 6363 Å lines are weak and in a region that is poorly cleaned of telluric contaminants.

2.5.3   Odd elements

Odd elements, Na and Sc, are listed in Table D.4. These have different nucleosynthetic sources and are not related to one another. We also include a comment on Li upper limits at the end of this section.

In metal-poor stars, sodium typically forms with the α-elements during core collapse SN. On the other hand, scandium forms in the iron core of a massive star with a yield that strongly depends on the proton-to-neutron ratio (Y_e), and it is very sensitive to neutrino processes (e.g., Woosley et al., 2002; Curtis et al., 2019).

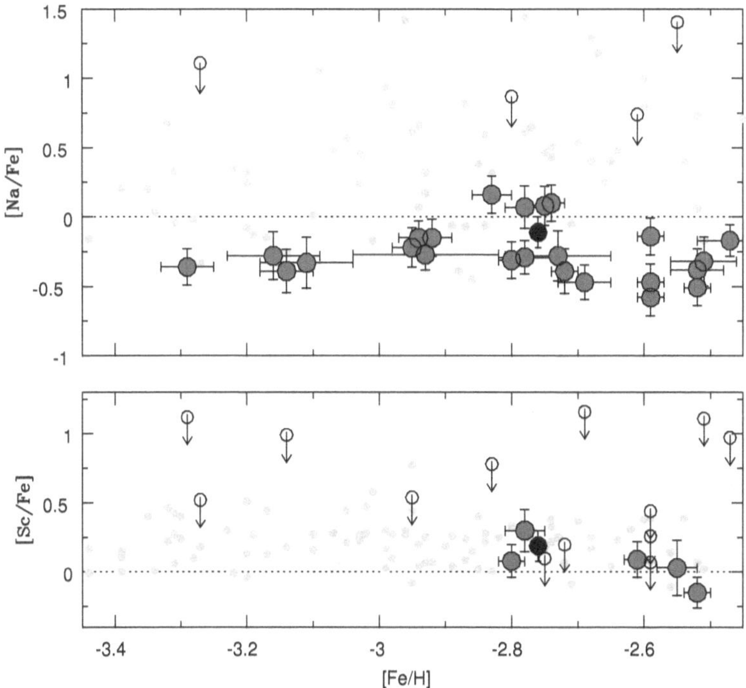

Figure 2.14 Na and Sc abundances and upper limits in the 28 new metal-poor stars ([Fe/H] < −2.5); symbols the same as in Fig. 2.11. NLTE corrections for Na have been applied from INSPECT (Amarsi et al., 2016).

Sodium abundances are initially from the LTE analysis of the Na I D lines (5889.9, 5895.9 Å), which unfortunately can be strong, therefore sensitive to microturbulence in a 1D LTE analysis, and also contaminated by interstellar Na. Furthermore, since they originate from the Na I ground state, they are subject to NLTE effects. NLTE corrections are similar between the INSPECT database and Mashonkina et al. (2017b); $[Na/H]_{NLTE} = [Na/H]_{LTE} + \Delta Na$, where $\Delta Na = -0.1$ to -0.6 dex. The Na I subordinate line (5688.2 Å) could only be used for upper limit estimates at the SNR of our spectra.

Despite the large NLTE corrections, four stars were found with initially very high Na I abundances (Pristine_251.4082+12.3657, Pristine_193.8390+11.4150, Pristine_217.5786+14.03 and Pristine_250.6963+08.3743, in order of decreasing metallicity). These four stars also have the lowest radial velocities in our sample (-5, +4, -16, and -4 $km\,s^{-1}$, in order of decreasing metallicity), and we suggest they are contaminated by the interstellar Na lines. To test this, their Na I D line shapes were compared with other spectral lines in the same stars and found to be slightly broader (occassionally, the line core is even split); their Na I D line shapes are also broader than similar stars with higher radial velocities (where the interstellar lines are often seen offset from the stellar lines). Thus, in Fig. 2.14, the highest Na abundances are noted as upper limits only since they are most likely blended, and for the other stars the NLTE corrected Na abundances are shown.

Sc II has been measured from 1-3 lines (5031.0, 5526.8, 5657.9 Å) in five metal-poor *Pristine* stars, and the comparison star HD 122563, and upper limits were determined in the others. With an odd number of nucleons, this species undergoes strong hyperfine splitting, which affects line formation through de-saturation. The HFS corrections were found to be small (< 0.1). Upper limits have also been determined for Sc II in most of the other new metal-poor stars. Upper limits were examined for Mn I as well, but did not provide interesting constraints.

Lastly, we mention Li in this section. Estimates from the Li I 6707 Å line provide upper limits that do not provide meaningful constraints, i.e., the upper limits are above the standard Big Bang nucleosynthesis value of A(Li) = 2.7 (e.g., from WMAP, Spergel et al., 2003). Only two stars (Pristine_229.1219+00.9089 and Pristine_237.8246+10.1426) have 3σ equivalent width (35 mÅ) upper limits of A(Li) $\leq$ 2.2, which is similar to most metal-poor stars that lie on (or below) the Spite Plateau (e.g., see Aguado et al., 2019a; Bonifacio et al., 2018).

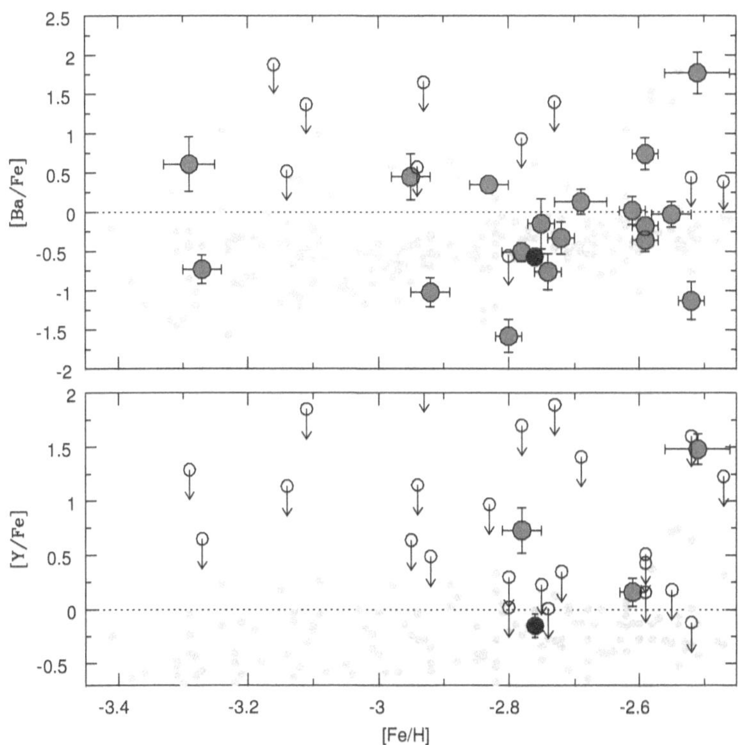

Figure 2.15 Ba and Y abundances and upper limits in the 28 new metal-poor stars ([Fe/H] < −2.5); symbols the same as in Fig. 2.11 with the exception of the Galactic comparison stars from Roederer et al. (2014, small grey points). We identify one star near [Fe/H]= −2.5 (Pristine_214.5556+07.4669) as an r-process rich star, significantly enriched in both Y and Ba.

2.5.4 Heavy elements

Abundances for the neutron-capture elements Y and Ba in the 28 new very metal-poor stars are listed in Table D.3. Up to four lines of Ba II (4554.0, 5853.7, 6141.7, and 6496.9 Å) and two lines of Y II (4883.7, 4990.1 Å) could be measured. Unfortunately no lines or useful upper limits for Eu are available in our CFHT spectra. When no lines were observable, we determined upper limits from 3σ minimum equivalent width estimates. Hyperfine splitting and the isotopic splitting has been included in the Ba analysis. Most stars have [Ba/Fe] in good agreement with the Galactic comparison stars.

All six lines were measured in only one star near [Fe/H]= -2.5 (Pristine_214.5555+07.4670). This star is enriched in both Y and Ba, and we identify it as an r-process rich star. Without Eu, it cannot be further classified as r-I or r-II (Christlieb et al., 2004; Sakari et al., 2018a). Studies of r-process rich stars have found a nearly identical main r-process pattern (from barium, A=56, to hafnium, A=72) in all types of stars, in all environments, and with variations only between the lightest and heaviest elements (see Roederer et al., 2010; Hill et al., 2017; Sakari et al., 2018b, and references therein). No other elements stand out in this star; however, as one of the hotter turn-off stars in our sample, there are not many other features or elements to analyse at the SNR of our spectra.

Two more stars show [Ba/Fe]$\gtrsim$+0.5 (Pristine_237.8246+10.1426, Pristine_210.0166+14.6289). These lie above the typical [Ba/Fe] values found in the Galactic halo metal-poor stars by Roederer et al. (2014), and their results are securely derived from 2 - 4 Ba II line measurements. However, no Y II lines were observed in either (and the Y II upper limits do not provide useful constraints). The two may be moderately r-process enriched stars.

Possibly of greater interest are the two most Ba-poor stars (Pristine_181.4395+01.6294 and Pristine_193.8390+11.4150). Low Ba is very unusual at their metallicities when compared with the other Galactic halo stars, as seen in Fig. 2.15. This composition is similar to stars in the Segue 1 and Hercules ultra faint dwarf (UFD) galaxies (Frebel et al., 2014; Koch et al., 2013). In Segue 1, the Ba-poor stars were discussed as representative of inhomogeneous enrichment by a single (or few) supernova events, and therefore possibly related to first stars. Higher SNR data for these two stars is warranted in order to test this hypothesis.

Finally, one star (Pristine_245.8356+13.8777) shows a high Y II abundance, but normal Ba II abundance. A similar star was recently studied by Caffau et al. (2019, J0222-0313), where the authors show it is a CEMP-s star, having undergone mass transfer in a binary system with an Asymptotic Giant Branch (AGB) star. However, they also suggest that the

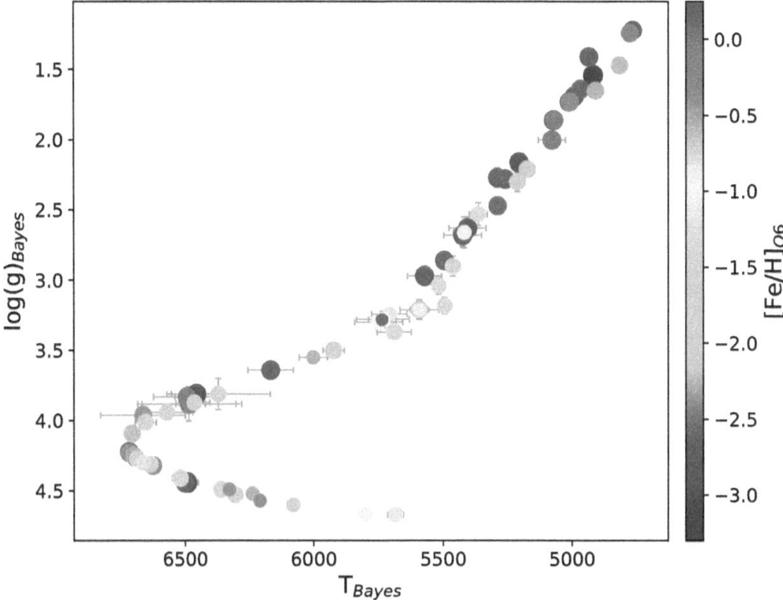

Figure 2.16 The HRD for the 70 metal-poor candidates in the *Pristine* survey, colour-coded by their ("Quick Six") metallicities $[Fe/H]_{Q6}$ as determined from our high resolution CFHT ESPaDOnS spectrum and Bayesian inference analysis. Stars that are not very metal-poor, with $[Fe/H] > -2.0$, are located over all stellar parameters.

AGB star in this system may have undergone a proton ingestion event just before the mass transfer that produced an enhancement in only the first s-process peak elements.

2.5.5   Abundance uncertainties

Total uncertainties in the chemical abundances are a combination of the measurement uncertainties and systematic errors in the stellar parameters, added in quadrature. For the measurement errors, when fewer than 4 lines are available for an element X, then we adopt the larger of $\sigma(X)$ or $\sigma(Fe\,\textsc{i})/sqrt(N_X)$. Since Fe $\textsc{i}$ lines are measured across the entire spectrum and over a range of equivalent widths and excitation potentials, then this assumes that $\sigma(Fe\,\textsc{i})$ captures the minimum measurement quality of our spectra. For the systematic errors, due to uncertainties in the stellar parameters, we determine the impact of the 1σ changes in temperature, gravity, and metallicity listed in Table D.2.

A sample of the systematic uncertainties for three stars that cover the parameter space of this sample are shown in Table D.6. It can be seen that temperature tends to be the dominant systematic error in the analysis of most elements. While we could further investigate the impact of the final metallicities [Fe/H] and uncertainties σ([Fe/H]) through iterations in the Bayesian inference method on the model atmospheres parameters, we did not; the only stars that we follow up in detail are those that did prove to be very metal-poor, therefore the impact of adjusting for the final metallicities on the other stellar parameters is very small.

2.6   Discussion

A total of 70 (out of 115) bright, metal-poor candidates have been observed with the CFHT ESPaDOnS spectrograph from the original footprint ($\sim$1000 sq deg) of the *Pristine* survey. These targets were selected to have a high probabilily for [Fe/H]$_{\text{Pristine}}$ < −2.5, when the *Pristine* CaHK filter was calibrated with the SDSS g-i and g-r colours (60 stars), or only the SDSS g-r colour alone (10 stars). We carry out a model atmospheres analysis by adopting stellar parameters determined from a Bayesian inference method that uses the SDSS colours, Gaia DR2 parallaxes, and MESA/MIST isochrones, assuming the initial *Pristine* survey metallicities. Out of these 70 selected stars, we have found 28 to indeed have low metallicities, [Fe/H]$\leq$ −2.5 (40%). The *Pristine* survey had also predicted 27 stars would have [Fe/H]$\leq$ −3.0, and 5 were found (19%). Of the 42 remaining stars (−2.5 < [Fe/H]$_{Q6}$ < +0.25), there are no obvious relationships with any other stellar parameters (e.g., see Fig. 2.16), although we notice that all of the candidates on the upper red giant branch were successfully selected and confirmed to be metal-poor stars.

The selections made in this chapter differ from those used by Youakim et al. (2017) and Aguado et al. (2019), see Section 2.2, being far more strict in the metal-poor probability cuts. Furthermore, about 1/3 of the targets in this program were observed before the selection criteria were finalized. Nevertheless, our success rates are very similar to the results from the medium resolution surveys. We do not reproduce the (lower) success rates for bright stars seen in earlier *Pristine* survey papers (Caffau et al., 2017; Bonifacio et al., 2019a), partially due to our improved (more strict) selection criteria, partially due to differences between the SDSS and APASS photometry, and possibly due to the larger number of stars in this sample.

In the remainder of this Discussion, I examine the kinematic and orbital properties of the 70 metal-poor candidates in this chapter, and correlate those with their chemical abundances. I caution that these calculations and our interpretations are highly dependent on the accuracy of the adopted Milky Way potential (described in the next section). For example, our orbit integrations do not account for effects like the Galactic bar, which can significantly influence halo star orbits (e.g., Price-Whelan et al., 2016; Hattori et al., 2016; Pearson et al., 2017).

2.6.1 Kinematics and Orbits

Galactocentric velocities (U, V, W) are calculated for each star from their Galactic Cartesian coordinates (X,Y,Z) following the methods of Bird et al. (2019). The distance between the Sun and the Galactic centre is taken to be 8.0 kpc, the Local Standard of Rest circular velocity is V_{circ} = 239 km s^{-1}, and the peculiar motion of the Sun is (U_0 = 11.10 km s^{-1}; V_0+ V_{circ} = 251.24 km s^{-1}; W_0 = 7.25 km s^{-1}, as described in Schönrich et al. (2010). The sign of U_0 is changed so that U is positive towards the Galactic anticentre. Errors in these velocities are propagated from the uncertainties in proper motion, radial velocities, and distance by calculating the mean dispersions from 1000 Monte-Carlo realisations, and selecting from a Gaussian distribution in each of the original quantities.

With the distances from the Bayesian inference analysis[11], precision radial velocities from our high resolution spectra, and proper motions from the Gaia DR2 database, then the orbital parameters for the sample are calculated using the Galpy package (Bovy, 2015). The *MWPotential14* is adopted, a Milky Way gravitational potential composed of a power-law, exponentially cutoff bulge, Miyamoto Nagai Potential disk, and Navarro et al. (1997) dark matter halo. A more massive halo is chosen following Sestito et al. (2019), with a mass of of 1.2 x 10^{12} $M_\odot$ which is more compatible with the value from Bland-Hawthorn & Gerhard (2016).

The UVW velocities for the 70 highly probable metal-poor stars in this sample are given in Table D.7. The Toomre diagram for these objects are shown in Fig. 2.17, colour-coded by the [Fe/H]$_{Q6}$ metallicities. Most of the metal-poor stars in our sample have halo-like velocities, as expected for their metallicities. One very metal-poor star (Pristine_183.6849+04.8619, discussed below) appears to have disk-like dynamics.

---

[11]For three stars, we reverted back to distances from their 1/parallax values based on unrealistical outer halo distances and other orbital properties. Two of these stars were discussed at the end of Section 2.4.1, and a third star is discussed in Appendix A

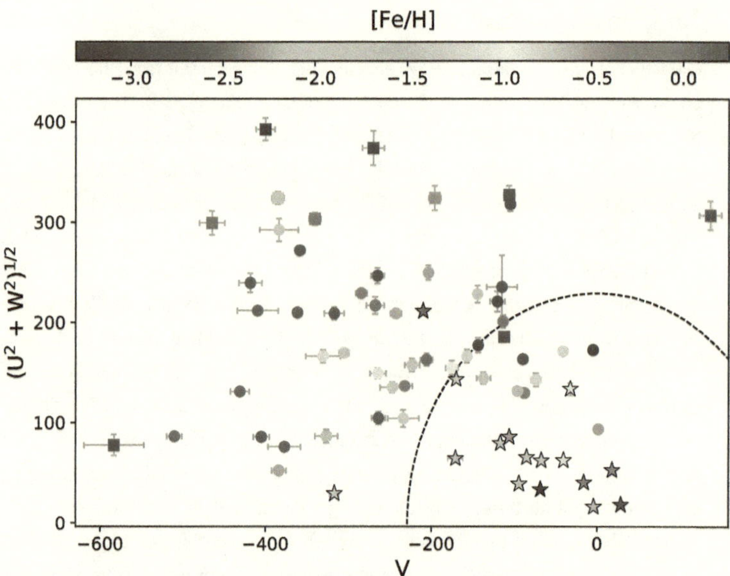

Figure 2.17 Toomre diagram for the 70 highly probable metal-poor stars in our *Pristine* survey sample. Dashed line represents stars potentially with disk dynamics (V_{circ} = 239 km s^{-1}). Symbols the same as in Fig. 2.18. Units are in km s^{-1}.

2.6.2   Orbit Analyses

To investigate the relationships between the chemical and kinematic properties of the stars in our sample, we examine their maximum excursions. This includes the apocentric and pericentric distances (R_{apo} and R_{peri}), perpendicular distance from the Galactic plane (Z_{max}), and eccentricity (e) of the derived orbits; see Table D.7.

In Fig. 2.18, stars with $R_{apo} < 15$ kpc and $Z_{max} < 3$ kpc are considered to be confined to the Galactic plane (16 stars), while stars with $R_{apo} > 30$ kpc are considered to be members of the outer halo (10 stars). The outer halo star Pristine_251.4082+12.3657 has the largest R_{apo} distance in our sample, with a highly eccentric orbit, and it is one of the most metal poor stars ([Fe/H]=−3.3), with low abundances of [Mg/Fe] and [Ca/Fe] (see Fig. 2.12), and also low [Ba/Fe]. This chemical signature is typical of stars in or accreted from the nearby dwarf galaxies. Alternatively, it may have been accreted from an *ultra* faint dwarf galaxy, since its chemistry is also similar to the unique stars CS 29498-043 and CS 29249-037 (Aoki et al., 2002; Depagne et al., 2002), both near [Fe/H]= −4. These stars have been proposed to be second-generation stars, that formed from gas enriched by a massive Population III first star, exploding as a fall-back supernova (see also Frebel et al., 2019), and as such they would have formed in a now accreted *ultra* faint dwarf galaxy.

In Fig. 2.19, only stars with $R_{apo} < 30$ kpc are shown. Clearly, most of the stars confined to the Galactic plane ($Z_{max} < 3$ kpc) are the relatively metal-rich (interloping) stars in our sample. However, one of the most metal-poor stars (Pristine_183.6849+04.8619, [Fe/H] = −3.1) is also confined to the Galactic plane with a nearly circular orbit (e=0.3). This was also seen in the Toomre diagram (Fig. 2.17). A detailed view of the orbit of this star is shown in Fig. 2.20. Most of the spectral lines in this star are weak and so we were unable to determine many elemental abundances, only [Mg/Fe]=+0.13 (±0.14) and [Na/Fe]=−0.18 (±0.17), which are both quite low for a typical halo metal-poor star. Ultra metal-poor stars ([Fe/H]< −4) have been found on similar quasi-circular and planar orbits by Sestito et al. (2019), and interpreted as stars that may have been brought in during the early merger phase of the building blocks of the proto-MW that eventually formed the disk.

Several (8) stars in our sample have orbits that take them deep into the Galactic bulge ($R_{peri} < 1$ kpc). All of these stars are on highly radial orbits (e > 0.8), and two are very metal-poor; Pristine_250.6963+08.3743 at [Fe/H]= −2.55 ± 0.03, and Pristine_201.8710+07.1810 at [Fe/H]= −2.93 ± 0.11. While the former star shows typical halo abundances in [(Mg,Ca,Ti)/Fe] = +0.4 (±0.4), the latter is clearly challenged in α-elements, [(Na,Mg)/Fe] = −0.1 (±0.2). It is difficult to discern whether these stars formed in the bulge

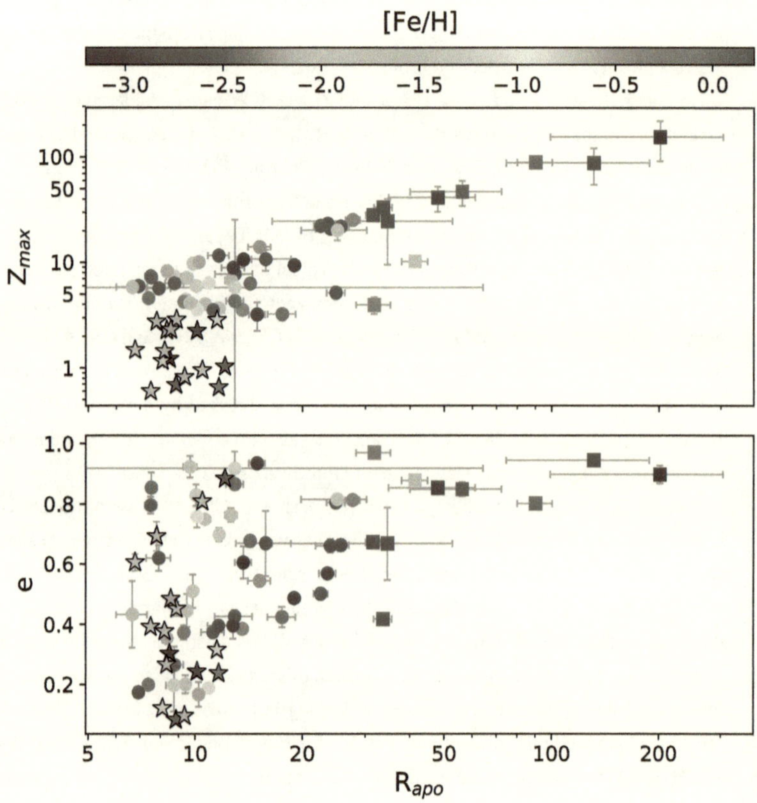

Figure 2.18 Perpendicular distance from the Galactic plane (Z_{max}) and eccentricity (e) of the orbits vs apocentric distance (R_{apo}) for the 70 high probability metal-poor stars in this chapter. For targets with $R_{apo} < 15$ and $Z_{max} < 3$ kpc I adopt "star" symbols, for $R_{apo} < 30$ kpc I adopt circle symbols, and when $R_{apo} > 30$ kpc I adopt square symbols. All targets are colour-coded by their [Fe/H]$_{Q6}$ metallicities.

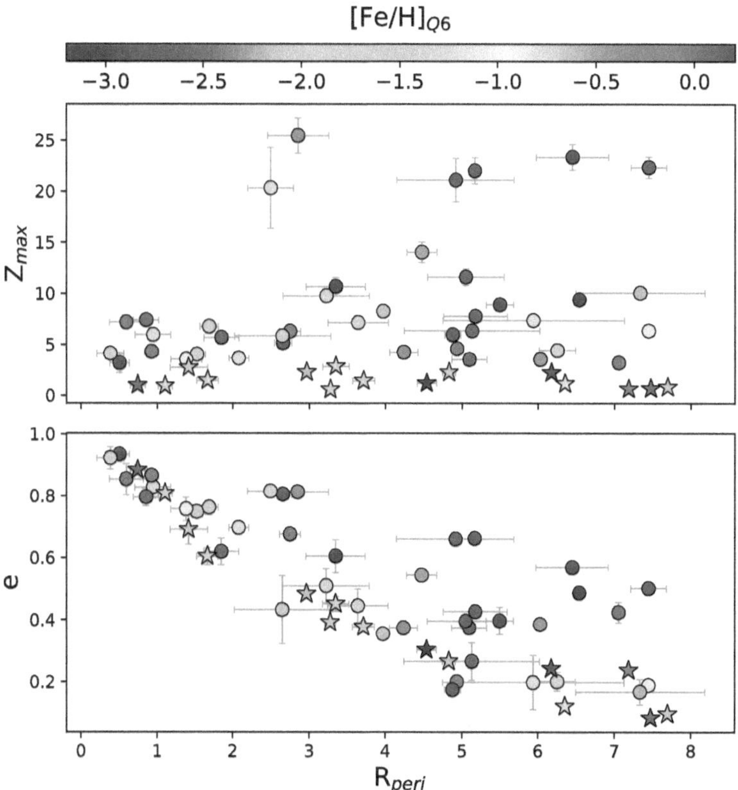

Figure 2.19 Z_{max} and eccentricity of the orbits vs pericentric distance R_{peri} for the stars within $R_{apo} < 30$ kpc. Symbols the same as in Fig. 2.18. The very metal-poor star confined within $Z_{max} = 1$ kpc (Pristine_183.6849+04.8619) near $R_{peri} = 4.5$ kpc can be seen more clearly in this plot than Fig. 2.18.

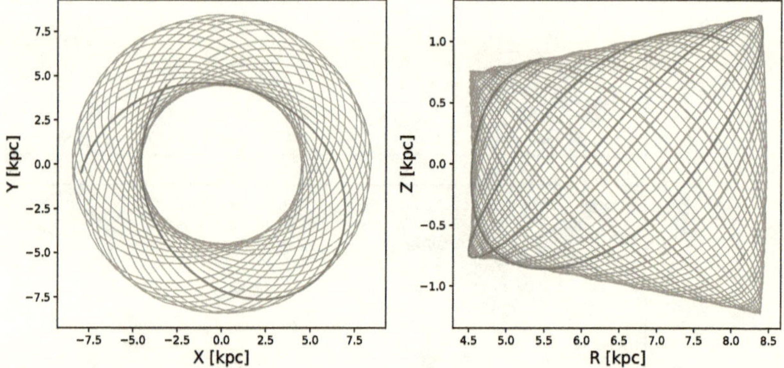

Figure 2.20 The orbit for the very metal-poor star Pristine_183.6849+04.8619, from our adopted Galactic potential. The orbital properties are R_{apo} = 8.5 kpc, Z_{max} = 1.2 kpc, and eccentricity e= 0.3. A sample single orbit is shown in red.

and have been flung out or if they have been accreted from the halo (or a dwarf galaxy) and moved inwards. As metal-poor stars in the bulge are thought to be older in absolute age (Tumlinson, 2010; Howes et al., 2016; Starkenburg et al., 2017b; El-Badry et al., 2018a; Frebel et al., 2019), then these could be extremely valuable objects for studies of the earliest stages of star formation in the Galaxy.

2.6.3 Action Parameters

The orbital energy (E) and action parameters (vertical J_z, azimuthal J_ϕ) were determined during the Galpy orbit calculations (discussed above); these are shown in Fig. 2.21 and provided in Table D.7. Values are scaled by the solar values, where $J_{\phi\odot} = 2009.92$ km s^{-1} kpc, $J_{z\odot} = 0.35$ km s^{-1} kpc and $E_\odot = $ -64943.61 (km s^{-1})^2. It is worth noting that stars with $J_\phi/J_{\phi\odot} = 1$ rotate like the Sun around the Galactic Centre.

The very metal-poor stars are roughly evenly distributed between retrograde and prograde orbits, i.e., between $-1 < J_\phi/J_{\phi\odot} < 1$. The most retrograde metal-poor star with a bound orbit (near $J_\phi/J_{\phi\odot} = -1$) is Pristine_198.5486+11.4123. This star has $Z_{max} = 3.2$ kpc, placing it very close to the Galactic plane. Therefore, this star is travelling at nearly the speed of the Sun but in the opposite direction, close to the Galactic plane. This orbit is certainly unusual and suggests it may have been accreted from a dwarf galaxy; however, its chemistry is like that of a normal metal-poor star, [Fe/H]= -2.5, [Mg/Fe]=+0.3, and [Ba/Fe]<+0.4.

The very metal-poor star Pristine_251.4082+12.3657, identified as having the largest R_{apo} value in this sample is also found to have a large $Jz/Jz_\odot$ value and an unbound orbit ($E/E_\odot < 0$). In total, three stars in Fig. 2.21 appear to have unbound orbits, although we caution that our uncertainties in their orbits are quite small when based on the very small distance errors from the Bayesian inference method. Examination of their parallax errors show that their orbits could be bound, consistent with $E/E_\odot \sim 0$. In Appendix A, I examine five more stars that appear to be dynamically unbound when their Bayesian inferred distances are used to determine their orbits. Two of those stars were discussed in Section 2.4.1, and it was shown that the orbital properties for these two metal-rich stars were significantly improved when 1/parallax was adopted for their distances. The same was found for a third star Pristine_213.7879+08.4232, even though this star has been confirmed to be metal-poor. The parallax errors for these three stars are all very small, and therefore we have adopted the 1/parallax distance for the orbital analysis of these three stars. Finally, two stars were removed from this kinematic analysis, Pristine_181.4395+01.6294

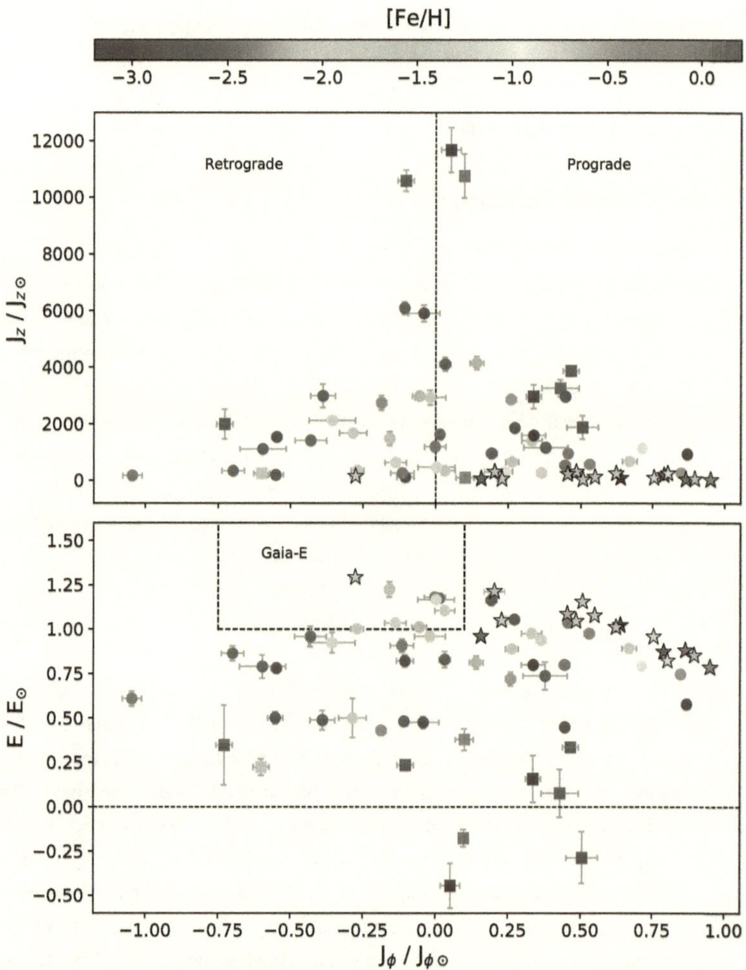

Figure 2.21 The orbit energies and rotational actions for the 70 high probability metal-poor stars in this chapter. The rotational action J_ϕ (= L_z) is compared with the vertical action J_z space (top panel) and the orbit energy (bottom panel), all normalized by the solar values. Prograde and retrograde regions are identified in the top panel. Symbols are as in Fig. 2.18. The region associated with Gaia-Enceladus is marked, above $E/E_\odot > 1$ and $-0.75 < J_\phi/J_{\phi\odot} < 0.1$ (Helmi et al., 2018; Haywood et al., 2018; Belokurov et al., 2018; Myeong et al., 2018, 2019).

and Pristine_182.5364+00.9431. Both stars have R_{apo} >500 kpc and e~1 resulting in extreme and unbound orbits for any distance adopted.

One of the most exciting discoveries from the Gaia DR2 dataset has been the identification of the Gaia-Enceladus dwarf galaxy (or galaxies) dissolved into the Milky Way halo. The region where stars may be associated with Gaia-Enceladus is shown in Fig. 2.21, i.e., $E/E_\odot > 1$ and $-0.75 < J_\phi/J_{\phi\odot} < 0.1$ (Helmi et al., 2018; Haywood et al., 2018; Belokurov et al., 2018; Myeong et al., 2018, 2019). This includes eight stars in our sample that range from $-2.5 < $ [Fe/H]$_{Q6} < -1.0$, with a mean metallicity of <[Fe/H]>=-2.0 ± 0.5; see Table D.7. Only one of these stars is sufficiently metal-poor to have made it into our detailed analysis sample, Pristine_250.6963+08.3743 ([Fe/H]=-2.55 ± 0.03). This star has high α-element abundances [(Ca,Ti)/Fe]~ +0.4, but lower magnesium such that [Mg/(Ca,Ti)]=-0.3, which is has been seen in some dwarf galaxies (e.g., Tri II, Venn et al. 2017). However, unlike most stars in dwarf galaxies, this star appears to have solar-like [Ba/Fe]~ 0 and [Sc/Fe]~ 0. It is unclear if this star is a true member of the original Gaia-Enceladus accretion event, but if so it would be amongst the most metal-poor stars yet found in that system (though also see Monty et al. 2020). As a final test, I examine the action-energy space of the newly discovered Gaia-Sequoia accretion event (Myeong et al., 2018, 2019), i.e., $E/E_\odot > 1$ and $J_\phi/J_{\phi\odot} < -1.5$, but find no targets in that parameter space.

2.7 Conclusions and Future Work

The results from our follow-up spectroscopy of 115 bright metal-poor candidates selected from the *Pristine* survey has been presented based on CFHT ESPaDOnS spectra. 28 new very metal-poor stars with [Fe/H]< -2.5 and five stars with [Fe/H]< -3.0 have been discovered, which imply success rates of 40% (28/70) and 19% (5/27), respectively. These rates are higher than previous surveys, though in line with the *Pristine* medium resolution programs. A detailed model atmospheres analysis for the 28 new very metal-poor stars, has provided stellar parameters and chemical abundances for 10 elements (Na, Mg, Ca, Sc, Ti, Cr, Fe, Ni, Y, Ba) and Li upper limits. Most stars show chemical abundance patterns that are similar to the normal metal-poor stars in the Galactic halo; however, we also report the discoveries of a new r-process rich star (Pristine_214.5556+07.4670), a new CEMP-s candidate with [Y/Ba]> 0 (Pristine_245.8356+13.8777), and a [Mg/Fe] challenged star (Pristine_251.4082+12.3657) which has an abundance pattern typical of stars in dwarf galaxies or, alternatively, gas enriched by a massive Population III first star exploding as a fall-back supernova. Two stars are also interesting because they are quite Ba-poor

(Pristine_181.4395+01.6294 and Pristine_193.8390+11.4150), and resemble stars in the Segue 1 and Hercules UFDs, which have been interpreted as evidence for inhomogeneous enrichment by a single (or few) supernovae events, and therefore possibly related to first stars.

The kinematics and orbits for all 70 of the metal-poor candidates have been determined using Gaia DR2 data, our radial velocities, and adopting the *MWPotential14* in the Galpy package (with a slightly more massive halo). The majority of the confirmed metal-poor stars are members of the Galactic halo, although some stars show unusual kinematics for their chemistry. We report the discovery of a very metal-poor ([Fe/H] = -3.2 ± 0.1) star (Pristine_183.6849+04.8619) with a slightly eccentric ($e = 0.3$) prograde orbit confined to the Galactic plane ($Z_{max} < 1.2$ kpc). We also find a metal-poor ([Fe/H] = -2.5 ± 0.1) star (Pristine_198.5486+11.4123) on a highly retrograde orbit (V= -510 km s^{-1}, $J_\phi/J_{\phi\odot} = -1.0$) that remains close to the Galactic plane ($Z_{max} < 3.2$ kpc). These two stars do not fit standard models for the formation of the Galactic plane, pointing towards more complex origins. An additional eight stars were found to have orbital energy and actions consistent with the Gaia-Enceladus accretion event, including one very metal-poor star (Pristine_ 250.6963+08.3743) with [Fe/H]=-2.5 and chemical abundances that are common for stars in dwarf galaxies. Finally, eight stars have highly radial orbits that take them deep into the Galactic bulge ($R_{peri} < 1$ kpc), including two very metal-poor stars (Pristine_250.6963+08.3743 at [Fe/H]= -2.55 ± 0.03, and Pristine_201.8710+07.1810 at [Fe/H]= -2.93 ± 0.11, the latter star is also low in α-elements). If these stars formed in the bulge, they could be extremely valuable for studies of the earliest conditions for star formation in the Galaxy.

Currently, I am running a Gemini/GRACES Large and Long Program to follow-up with high SNR (> 100) spectra for our best metal-poor candidates ([Fe/H] < -3.5) and with V < 17 selected from medium resolution spectroscopy (see Chapter 3). We also plan to observe a selection of these stars with the upcoming Gemini GHOST spectrograph (Chene et al., 2014; Sheinis et al., 2017), which is anticipated to have excellent throughput at blue-UV wavelengths, providing far more iron-group lines for stellar parameter assessments and many more spectral lines of heavy neutron-capture (and light) elements.

In the near future, massively multiplexed high resolution spectroscopic surveys (R > 20,000) will be initiated, including the European WEAVE survey at the Isaac Newton Telescopes (Dalton et al., 2012), the 4MOST survey at ESO (De Jong et al., 2019), and the SDSS-V survey comprising fields in both the northern and southern hemispheres (Kollmeier et al., 2017). These will provide the truly large statistical samples needed for studies of the metal-poor Galaxy.

Chapter 3

The *Pristine* Survey: Gemini-GRACES chemo-dynamical study of newly discovered very metal-poor stars in the Galaxy

The following work has been submitted to MNRAS at the time of this writing. Authors: Collin Kielty, Kim Venn, Federico Sestito Else Starkenburg, Nicolas F. Martin, David S. Aguado, Anke Arentsen, Sébastien Fabbro, Jonay I. González Hernández, Vanessa Hill, Pascale Jablonka, Carmela Lardo, Lyudmila I. Mashonkina, Julio F. Navarro, Chris Sneden, Guillaume F. Thomas, Kris Youakim, Spencer Bialek, Rubén Sánchez-Janssen

Personal Contributions

Already stated in Section 1.2.1, the data presented in this chapter were collected over four semesters at Gemini-N. The 2018A data comes from a Gemini/GRACES campaign that I led as P.I., while the data from the remaining semesters come from an ongoing Gemini/GRACES Large and Long Program (LLP) (P.I. Kim Venn). I have taken over a large majority of the LLP duties, selecting the high priority targets from the *Pristine*MRS sample, preparing the observing and instrument configurations required, and writing a substantial amount of code for a data reduction suite to process the GRACES data from the 2018B-2019B semesters. All figures and analysis work was completed by myself with the advice of others.

3.1 Introduction

The oldest and most metal-poor stars contain a fossil record of the star formation processes in the early Universe. These first stars would have been composed solely of hydrogen, helium, and trace amounts of lithium (Steigman, 2007; Cyburt et al., 2016); without metals to efficiently cool the gas, large Jeans masses, and thereby very massive stars ($M_* \gtrsim 100M_\odot$) are predicted to have formed (Silk, 1983; Tegmark et al., 1997; Bromm et al., 1999; Abel et al., 2002; Yoshida et al., 2006). In addition, improved gas fragmentation models (e.g., Clark et al., 2011a; Schneider et al., 2012) and the discovery of very low mass ultra metal-poor stars ([Fe/H]< −4; Keller et al. 2014; Starkenburg et al. 2017a; Schlaufman et al. 2018; Nordlander et al. 2019) have suggested that lower mass, long lived stars may have also formed in these pristine environments (e.g., Nakamura & Umemura, 2001; Wise et al., 2012). Notably, if ≤ 0.8 $M_\odot$ stars were to form, they could still exist today on the main sequence, and are expected to have undergone very little surface chemical evolution over their lifetimes.

Detailed studies could provide invaluable constraints on (1) the nucleosynthetic yields from the first stars and the earliest supernovae (e.g., Heger & Woosley, 2010; Ishigaki et al., 2014; Tominaga et al., 2014; Clarkson et al., 2018; Nordlander et al., 2019), (2) the physical conditions in the high redshift universe (where stars are too faint to be observed individually; Tegmark et al., 1997; Freeman & Bland-Hawthorn, 2002; Cooke & Madau, 2014; Hartwig et al., 2018; Salvadori et al., 2019), (3) the primordial initial mass function (e.g., Greif et al., 2012; Susa et al., 2014), and (4) early Galactic chemical evolution processes (see Sneden et al., 2008; Tolstoy et al., 2009; Roederer et al., 2014; Yoon et al., 2016; Wanajo et al., 2018; Kobayashi et al., 2020, and references therein).

Dedicated searches for old, metal-poor stars in the Milky Way and in its dwarf galaxy satellites were initiated over two decades ago (e.g., Bond, 1980; Carney & Peterson, 1981; Beers et al., 1985, 1999). Chemical abundances have now been measured for hundreds of extremely metal-poor stars (EMP), where $[Fe/H]$[1] ≤ -3 (e.g., Aoki et al., 2013; Yong et al., 2013; Cohen et al., 2013; Frebel & Norris, 2015; Matsuno et al., 2017; Aguado et al., 2019b) and over a dozen stars with $[Fe/H] < -4$ (see Aguado et al., 2018a,b; Bonifacio et al., 2018; Starkenburg et al., 2017b; Aguado et al., 2019b; Frebel et al., 2019; Nordlander et al., 2019, and references therein). Most of the Galactic EMP stars show enhanced $[\alpha/Fe]$ abundances and a diversity of neutron capture element ratios, interpreted as the yields from core-collapse supernovae with different progenitor masses and explosion prescriptions. More recently, the importance of compact binary mergers to the r-process abundance ratios has also been revealed (Ji et al., 2016a; Roederer et al., 2016; Hansen et al., 2017). In dwarf galaxies, the abundance ratios of hydrostatic elements (*e.g.* O, Na, Mg), explosive elements (*e.g.* Si, K, Mn), complex elements (like Zn), and heavy elements (as Ba, Y) tend to be lower than those of their Galactic counterparts of similar metallicity. This reflects a range of differences in their star formation histories, including interstellar mixing efficiencies, star formation efficiencies, star formation rates, and effective mass functions of the dwarf galaxy systems (Venn et al., 2004; Tolstoy et al., 2009; Nissen & Schuster, 2010; McWilliam et al., 2013; Frebel & Norris, 2015; Hasselquist et al., 2017; Hayes et al., 2018; Lucchesi et al., 2020). The diversity of chemical abundance profiles seen in this sparse population of objects makes for challenging statistical studies. Targeted programs are needed to uncover larger samples of these (nearly) pristine stars.

The present day locations of metal-poor stars also a diagnostic of early galaxy formation. Based on cosmological simulations of the Local Group, it is thought that the Galactic

---

[1]We adopt the square bracket notation for chemical abundances relative to the Sun, such that $[X/Y] = \log(X/Y)_\odot - \log(X/Y)_*$, where X and Y are column densities for two given elements.

halo was formed through the accretion and disruption of dwarf galaxies at early epochs. Consequently, properties of the old, metal-poor stars seen in the halo manifest in the properties of their dwarf galaxy hosts (e.g., Ibata et al., 1994; Helmi et al., 1999; Johnston et al., 2008; Wise et al., 2012; Starkenburg et al., 2017b; El-Badry et al., 2018a). The arrival of precision proper motions from *Gaia* DR2 (Gaia Collaboration et al., 2018), and increasingly large datasets of stars with precision radial velocities from high-resolution spectroscopy, has enabled the determination of orbits for halo stars. The majority of the EMP halo stars have been shown to have high-velocities and eccentric orbits, consistent with acccretion from a dwarf galaxy (Sestito et al., 2019, 2020a; Di Matteo et al., 2020; Cordoni et al., 2020). A large population of halo stars has also been found with retrograde orbits and kinematics consistent with a halo merger remnant, e.g., *Gaia-Enceladus/Sausage* (Meza et al., 2005; Belokurov et al., 2018; Helmi et al., 2018; Myeong et al., 2018) and *Gaia-Sequoia* (Barbá et al., 2019; Myeong et al., 2019; Matsuno et al., 2017; Monty et al., 2020; Cordoni et al., 2020).

The Galactic halo is not the only place to look for old stars. Curiously, some EMP stars have been found on nearly circular, planar orbits (Caffau et al., 2012b; Sestito et al., 2019, 2020a; Schlaufman et al., 2018; Venn et al., 2020). Since the Galactic plane is thought to have formed ~ 10 Gyr ago (Casagrande et al., 2016b), these stars challenge the idea that the metal-poor stars are amongst the oldest stars accreted from the mergers of dwarf galaxies. The Galactic bulge is another environment to look for relics of first stars (White & Springel, 2000; Guo et al., 2010; Starkenburg et al., 2017b; El-Badry et al., 2018a). To date, ~ 2000 very metal-poor stars ([Fe/H] < -2.0) associated with the bulge have been found (García Pérez et al., 2015; Howes et al., 2015, 2016; Lamb et al., 2017; Arentsen et al., 2020a). Detailed chemical abundance analyses for these objects are limited, especially at the lowest metallicities, but they indicate that many of the metal-poor bulge candidates are chemically similar to halo stars (Lucey et al., 2019). In fact, estimates of their orbital properties using the exquisite astrometry from Gaia DR2 (Gaia Collaboration et al., 2018) suggests that up to 50% of these stars may be normal halo stars with highly elliptical and plunging orbits, a fraction which increases with decreasing metallicity (Lucey et al., 2020). However, while *Gaia* DR2 proper motions are extremely valuable in eliminating foreground metal-poor non-bulge members, the uncertainties on the parallax measurements for most stars in the bulge are currently too large for reliable distance estimates.

Regardless of where metal-poor stars are found, the union of stellar dynamics and chemical cartography provides a foundation for studies of *Galactic Archaeology* (e.g., Carney et al., 2003; Freeman & Bland-Hawthorn, 2002; Venn et al., 2004; Hayden et al.,

2015; Bovy et al., 2016; Hasselquist et al., 2017). Dedicated surveys have been successful in finding most of the metal-poor stars; these include the HK and Hamburg-ESO surveys (Beers et al., 1992; Christlieb et al., 2002b; Beers & Christlieb, 2005), the SDSS SEGUE, BOSS, and APOGEE surveys (Yanny et al., 2009; Eisenstein et al., 2011; Majewski & SDSS-III/APOGEE Collaboration, 2014), LAMOST (Cui et al., 2012), and more recently the narrow band imaging surveys *SkyMapper* (Keller et al., 2007; DaCosta et al., 2019; Onken et al., 2020) and *Pristine* (Starkenburg et al., 2017a). The discovery of more EMP stars is necessary to build a statistically significant sample to study the metal-poor Galaxy, and also to overcome cosmic variance in the chemo-dynamic analysis of stellar populations within it.

In this study, 30 new candidate EMP stars have been selected from the *Pristine* survey for follow-up high-resolution spectroscopy. *Pristine* is based on a MegaPrime/MegaCam imaging survey at the 3.6-metre Canada France Hawaii Telescope, using a unique narrow-band filter centered on the Ca II H & K spectral lines (*CaHK*). When *CaHK* is combined with the broad-band SDSS *gri* photometry (York et al., 2000) or Gaia DR2 photometry, this *CaHK* filter can been calibrated to find metal-poor candidates with $4000 < T_{eff} < 7000$ K (Youakim et al., 2017; Starkenburg et al., 2017a; Arentsen et al., 2020b). The *Pristine* survey has been shown to successfully predict stellar metallicities. $\sim 56\%$ of stars with [Fe/H]$_{Phot.} < -2.5$ are confirmed to have [Fe/H]< -2.5 ($\sim 23\%$ when [Fe/H]$_{Phot.} < -3.0$) based on follow-up spectroscopic studies (Youakim et al., 2017; Caffau et al., 2017; Bonifacio et al., 2019a; Aguado et al., 2019b; Venn et al., 2020). The target stars for this study were selected from the medium-resolution spectroscopic follow-up studies of EMP candidates found within the initial $\sim$2500 sq. degree footprint[2] of the *Pristine* survey (Martin et al., in prep.). In Section 4, we improve upon the the previous stellar parameter determinations; in Section 5, we calculate the chemical abundances or upper-limits for $\sim$15 elements; and in Section 6, we estimate the orbits using the Gaia DR2 database for a chemo-dynamical analysis. Together, these properties permit a study of the origins of these metal-poor stars themselves, ultimately to unravel the formation history of the Milky Way Galaxy.

3.2   Target Selection

Targets in this study have been selected from the original 2500 sq. degree *Pristine* photometric survey (Starkenburg et al., 2017a; Youakim et al., 2017) and medium-resolution

---

[2]We note that the *Pristine* survey has more than doubled its survey area since its initial footprint.

spectroscopic follow-up survey (Youakim et al., 2017; Aguado et al., 2019b). All of the medium-resolution spectra (hereafter MRS) were observed at the Isaac Newton Telescope (INT), and were analysed using the FERRE data analysis pipeline (Allende Prieto et al., 2006) and the ASSET grid of synthetic stellar spectra as described and published by Aguado et al. 2017b. FERRE searches for the best fit to the observed spectrum by simultaneously deriving the main stellar atmospheric parameters (effective temperature, surface gravity, and metallicity [Fe/H]), and it also determines the carbonicity, [C/Fe]. Uncertainties are found via Markov Chain Monte Carlo (MCMC) sampling around the best fit stellar parameters. Stars with a very high probability ($> 90\%$) for a *Pristine* metallicity $[Fe/H]_{phot} < -2.75$ in both the $(g - i)$ and $(g - r)$ calibrations are shown in Fig. 3.1 (in grey). The metallicities from the INT medium-resolution spectral analyses by FERRE for those *same stars* are also shown in Fig. 3.1 (in red). Clearly some of our initial *Pristine* metallicities were too low; however, significant improvements in the *Pristine* selection function have been made since our original MRS spectroscopic follow-up programs, in anticipation of providing targets for the WEAVE survey (Dalton et al., 2018).

Targets in this study with high-resolution spectroscopy have been selected with $[Fe/H]_{MRS} < -3$ and low temperatures ($T_{eff} < 6500\,K$) in the magnitude range $14.8 < V < 16.4$. These are noted in blue in Fig. 3.1. We have observed $<5\%$ of the full sample from the original footprint area, showing that a spectroscopic survey like WEAVE is necessary to reach all of them. Some stars with enhanced carbon were included ($[C/Fe]_{FERRE} > 1$); however, stars with non-normal carbon abundances were not given a higher priority in the target selection because the carbon abundances are typically unreliable when low SNR (< 25) and low-resolution data are analysed with the FERRE pipeline (see below, also Aguado et al., 2019b, Arentsen et al. in prep.).

The targets analysed in this study are listed in Table D.8. The stellar identifications (IDs) are a combination of their SDSS RA and DEC coordinates and V are in observer magnitudes calculated from SDSS u, g, r, i, z given the calibration by (Lupton et al., 2005). The other SDSS and *Pristine* filter magnitudes have been dereddened using E(B-V) from (Schlegel et al., 1998). Table D.8 also lists the information on the exposures and the SNR of the final combined spectrum for each target. Stellar parameters from the *Pristine* survey are listed in Table D.9. This includes the photometric *Pristine* metallicities ($[Fe/H]_{phot}$) and the colour temperatures (T_{phot}), which are an average of the dwarf and giant solutions from SDSS *gri* and *Pristine CaHK* photometry (see Starkenburg et al., 2017a). Table D.9 also includes the stellar parameters from the FERRE analysis of the INT medium-resolution spectroscopic survey by Aguado et al. (2019b) for our sample. The average SNR = 28 for

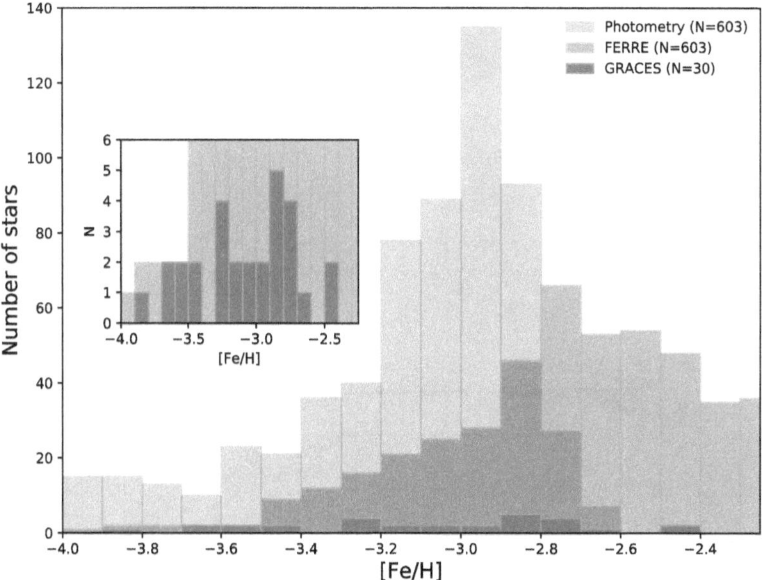

Figure 3.1 The metallicity distribution for the full sample of metal-poor stars found in the original *Pristine* footprint. The grey and pink distributions are for the same sample of stars (V<18, Aguado et al., 2019b), showing that medium-resolution spectral follow-up found slightly higher metallicities for this sample. GRACES metallicities from this study are shown in blue (inset included for higher detail). Clearly, there are many more EMP candidates available for high-resolution spectroscopic analyses.

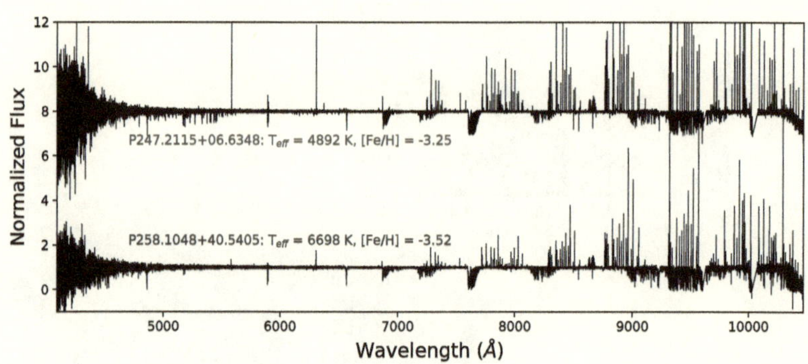

Figure 3.2 Full Gemini/GRACES spectra for the hottest and coolest stars in our sample. Even though the star+sky mode was used, the sky subtraction is clearly imperfect due to the presence of emission lines at longer wavelengths. The poor throughput below ~4800Å of the long 270-metre optical fibre can be seen as an decrease in the SNR at blue wavelengths.

the MRS of our targets, with a range $7 < SNR < 79$.[3] As mentioned above, the FERRE pipeline simultaneously determines the effective temperature (T_{eff}), surface gravity (log g), metallicity [Fe/H]$_{MRS}$, and carbonicity [C/Fe]$_{MRS}$ for each star. Radial velocities are also derived from MRS ($\Delta RV \sim 1$ km s^{-1}). Aguado et al. (2019b) investigated the FERRE carbonicity measurements as functions of both surface gravity and SNR. They find that carbon abundances are more reliably determined (systematic uncertainties of ~ 0.2 dex) for cool, lower gravity giants, primarily due to the increased strength of spectral lines at lower temperatures. Likewise, increased line strength means lower SNR is needed for a detection of the G band. Conversely, the uncertainties in the carbon abundances increases up to $\sim$0.6 dex for the hotter, higher gravity stars, where weaker spectral features and less reliable surface gravities are expected. Since the carbon measurement is so heavily dependent on both SNR and gravity, we do not report individual values for [C/Fe]$_{MRS}$ in Table D.9, and instead flag stars as C-rich candidates if [C/Fe]$_{MRS} > 1.0$. The stars are flagged further to note whether the carbon abundances are *reliable*, based on the SNR of the medium-resolution INT spectra analysed by FERRE.

3.3   Gemini-GRACES Observations

High-resolution spectra have been collected for 30 extremely metal-poor candidates with the Gemini Remote Access to CFHT ESPaDOnS Spectrograph (GRACES; Chene et al., 2014; Pazder et al., 2014). A 270-metre optical fibre links the Cassegrain focus at the Gemini-North telescope to the Canada-France-Hawaii Telescope ESPaDOnS spectrograph (a cross-dispersed high-resolution échelle spectrograph that covers the whole visible spectrum; Donati et al., 2006). We note that the exposure times we used were in good agreement with the GRACES ITC provided by Gemini Observatory. In the 2-fibre (object+sky) mode, spectra are obtained with resolution R~65,000; however, light below ~4800 Å is severely limited by poor transmission through the optical fibre link.

The GRACES spectra have been reduced using the Gemini "Open-source Pipeline for ESPaDOnS Reduction and Analysis" tool (OPERA, Martioli et al., 2012). This includes

---

[3]P258.1048+40.5405, the faintest star in the sample at $V = 16.06$, was observed for over 8.5 hours to reach a SNR high enough for a detailed chemical abundance analysis. The original MRS analysis for this object that was used in the target selection indicated a metallicity near [Fe/H]~ -5.0, warranting the long exposure time as this could be one of the most metal-poor stars known. Unfortunatley, as this star was being observed with GRACES, the MRS analysis for this object was re-done and a new, higher metallicity of [Fe/H]$= -3.76 \pm 0.73$ was measured. Though only Fe upper-limits could be placed for P258.1048+40.5405 from the GRACES spectra ([Fe/H]< -3.52, see Sec. 3.5.1), we do not suspect this star is much more metal-poor than [Fe/H]< -3.5 based on [Mg/H]$= -3.21$ and [Ca I/H]< -2.91.

standard calibrations (images are bias subtracted, flat fielded, extracted, wavelength calibrated, and heliocentric corrected). Starting from the individually extracted and normalized échelle orders (the *i.fits files), one continuous spectrum is stitched together by weighting the overlapping wavelength regions by their error spectrum, and co-adding as a weighted average. All visits for a given star are examined for radial velocity (RV) variations, potentially indicating binarity; no evidence for binary systems was found based on the criteria $\Delta RV \leq 1 \, km \, s^{-1}$, although we note radial velocity variations are unlikely to be measured due to the short cadence of our observations. All visits for a given star have then been co-added via a weighted mean using the error spectrum. The coadded spectrum has then been radial velocity corrected through cross-correlation with a high SNR comparison spectrum of the metal-poor standard star HD 122563. Other radial velocity standards were examined (e.g., Arcturus and a synthetic spectrum for a typical metal-poor RGB), however the solutions had the smallest uncertainties ($\sigma RV \leq 0.2 \, km \, s^{-1}$) when compared to HD 122563. As a final step, the RV corrected spectrum were re-normalized using k-sigma clipping with a nonlinear filter (a combination of a median and a boxcar). The effective scale length of the filter ranged from 8 to 15 Å, dependent on the crowding of the spectral lines, which was sufficient to follow the continuum when used in conjunction with iterative clipping. The full wavelength coverage of the final spectra for two sample stars is shown in Fig. 3.2, including the imperfect sky subtraction, as seen by the night sky emission lines above ~8000 Å.

3.4 Stellar Parameters and Radial Velocities

Stellar parameters (T_{eff} and $\log g$) have been determined using the Bayesian inference method described by Sestito et al. (2019). In this method, the SDSS *gri* magnitudes, Gaia DR2 parallax, *Pristine* survey metallicity, and MIST/MESA isochrones are used to determine the stellar temperature and surface gravity (T_{eff} and $\log g$; see Table D.10) for each star. Briefly, a probability distribution function of the heliocentric distance to each star is determined by combining its photometric and astrometric data with stellar isochrones within a Milky Way stellar density prior. Isochrones are taken from the MESA/MIST library (Paxton et al., 2011; Dotter, 2016; Choi et al., 2016), which reach the lowest metallicities ([Fe/H] ≤ -4), assuming a flat age prior between 11 and 14 Gyr (or $\log(A/yr) = 10.05$ to 10.15). Distances are calculated from the parallax, when the relative parallax errors are < 2%; also, a zero point offset on the parallax of -0.029 mas was adopted (Lindegren et al., 2018). The final stellar parameter, microturbulence ξ, was determined from the surface gravity values, using the calibrations by Sitnova et al. (2015) for dwarf stars and Mashonkina

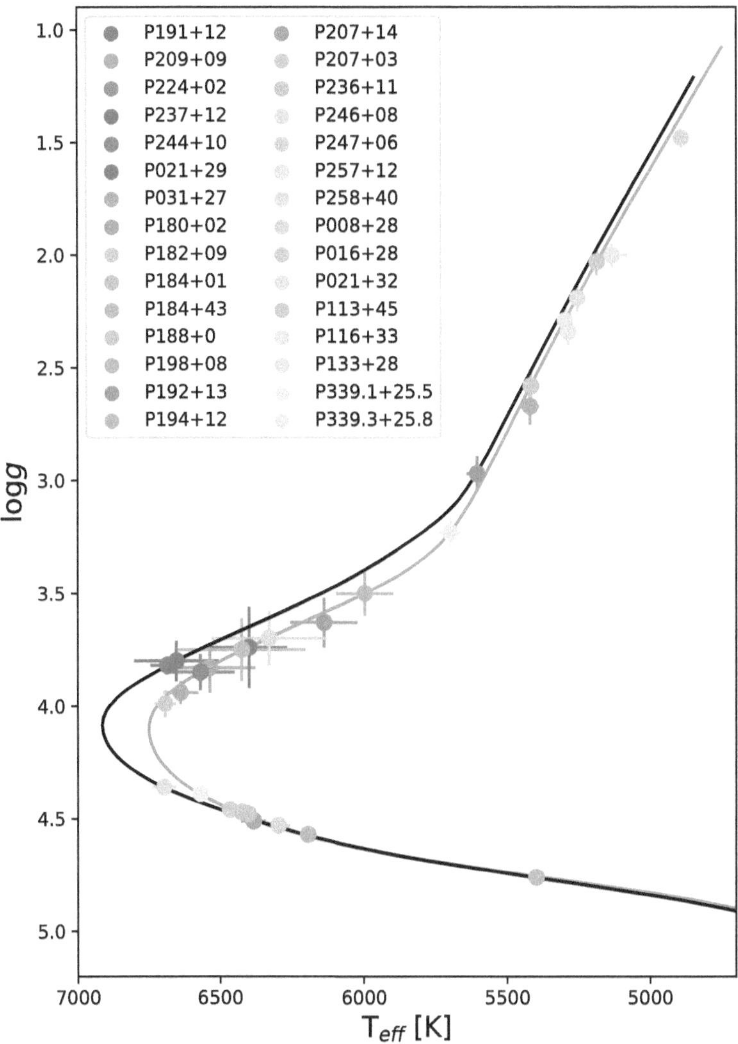

Figure 3.3 T_{eff} and log g inferred from the Bayesian inference method. MIST/MESA isochrones are [Fe/H] = -4.0 (black line) and [Fe/H] = -3.5 (grey line) for age = 14 Gyr.

et al. (2017c) for giants.

The targets in this study are shown in Fig. 3.3. A comparison of the stellar parameters determined from the Bayesian inference method (Table D.10) to those determined from the medium-resolution FERRE analysis (Table D.9) are shown in the bottom row of Fig. 3.4. The FERRE stellar parameter errors are the FERRE MCMC fitting errors to the spectra. The photometric colour estimates for temperature (Table D.8) are also shown for comparisons in the top row of Fig. 3.4, where separate dwarf and giant solutions have been averaged together. While the photometric temperatures tend to be hotter than the FERRE MRS temperatures, those same stars in FERRE tend to be cooler than determined in the Bayesian analysis. Thus, the photometric and Bayesian temperature results are in good agreement; typically $\Delta T_{eff} \sim 200$ K, but stars cooler than T=6000 K show hotter photometric temperatures, while stars hotter than T=6000 K show cooler temperatures. Due to the low SNR of the MRS, FERRE is forced to make a dwarf/giant distinction, forcing the gravities to values near the edge of the FERRE grid ($\log g$= 1.0 or 5.0). Similar trends in gravity have been seen when comparing the FERRE MRS results to other high-resolution spectral analyses of EMP stars in our group (Bonifacio et al., 2019b; Venn et al., 2020; Arentsen et al., 2020a).

The quality of the spectral parameters using the MIST/MESA isochrones for metal-poor stars has been questioned. Monty et al. (2020) found good agreement between spectroscopic and isochrone-mapping temperature determinations for stars with [Fe/H]> −2. However, for lower metallicities, the temperatures determined from isochrone-mapping tended to be hotter, by up to ΔT_{eff}=+500 K when [Fe/H]=−3.5. A similar result has been seen by Joyce & Chaboyer (2015, 2018) due to a range of (optimized) values for the convective mixing length parameter. We investigate this effect using three representative metal-poor stars from our sample (all with [Fe/H]∼ −3): a hot dwarf (P207.9290+03.2767), a cool giant (P236.9604+11.6155), and a cool dwarf (P184.2997+43.4721). All three stars have at least 10 Fe I lines, over a range of excitation potentials, allowing for a crude estimate from the slope in the line abundances versus excitation potential. A grid of models were run for each star, with offsets in T_{eff}(and $\log g$). Similar to the aforementioned temperature discrepancies seen in Monty et al. (2020), we find that lower temperatures (T_{eff}−500 K) produce the flattest slopes in χ vs. A(Fe I). Small offsets in gravity ($\log g$−0.5) would also bring the slightly revised (higher) Fe I metallicities into better agreement with Fe II. Thus, we confirm that the metal-poor MIST/MESA isochrones may have systematic offsets that may be affecting the accuracy of our stellar parameters using the Bayesian inference method. Yong et al. (2013) used the Y^2 isochrones (Demarque et al., 2004) to determine surface gravities for their program stars and did not observe notable offsets between their

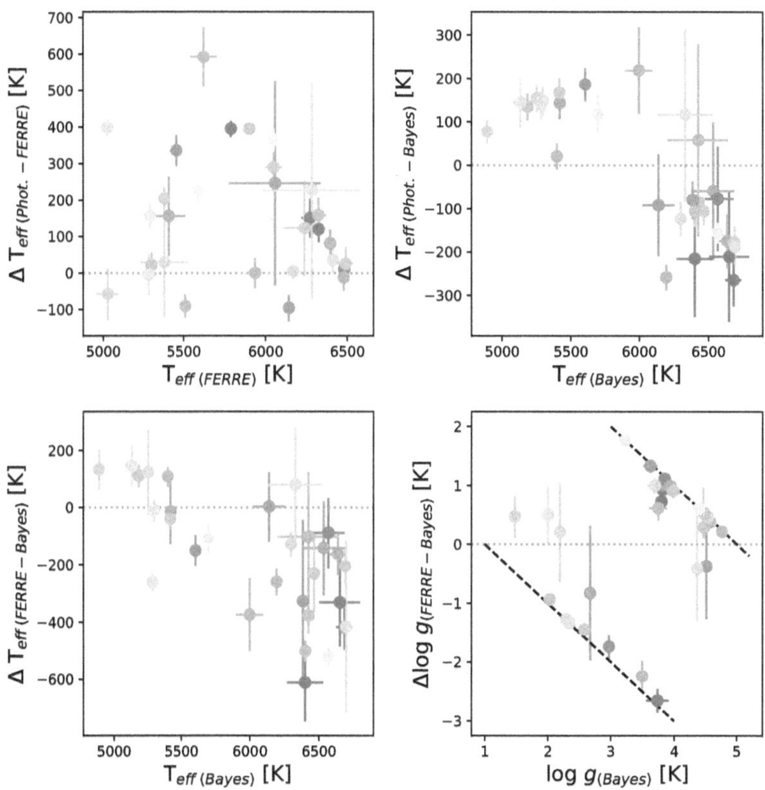

Figure 3.4 Comparison of T_{eff} and $\log g$ derived from photometry, medium-resolution spectroscopy, and a Bayesian analysis utilizing Gaia motions. Dashed black lines of fixed $\log g_{FERRE} = 1.0$ and 5.0 (the bounds of the FERRE grid) are shown in the bottom right panel. See Fig. 2 for star labels.

stellar parameters and those reported in the literature. We leave the investigation of stellar parameters determined from other isochrone models to future studies.

Based on the tests above, we acknowledge a potential systematic error in our stellar parameters; however, we do not adjust our stellar parameters in this chapter. This is because we do not have a sufficient number of well observed Fe I lines for a more accurate spectroscopic temperature measurement. We suspect part of these systematics arise from the fact that the MIST/MESA isochrones are only computed for scaled-solar composition at low metallicity. While this would likely affect the effective temperature derived from the isochrones, we do not expect this could fully explain the 500K offsets observed in some of our stars.

A comparison of radial velocity values from the medium-resolution spectra and our high-resolution GRACES spectra is shown in Fig. 3.5. There has been significant improvement in the radial velocity determinations between our initial MRS results (in *Pristine* VI, Aguado et al., 2019b) and the most recent MRS analysis (in *Pristine* X, Sestito et al., 2020b). The most recent MRS radial velocities show a mean offset of only $\Delta(RV) \sim 6$ km/s, with a dispersion of 12 km/s, relative to our HRS values.

3.5   Chemical Abundances

Chemical abundances are determined for each star using the stellar parameters discussed above and a classical model atmospheres analysis. Model atmospheres were generated using the most up-to-date models on the MARCS website[4] (Gustafsson et al., 2008, with additions by B. Plez); the OSMARCS spherical models are used when $\log g < 3.5$. Spectral lines of iron were selected from the recent metal-poor stars analyses by Norris et al. (2017) and Monty et al. (2020). Atomic data was adopted from the *linemake*[5] atomic and molecular line database. Chemical abundances are compared to the Sun using the Asplund et al. (2009) solar data.

Elemental abundances are calculated in a three step process: (1) the list of well known iron lines was examined in each spectrum for an initial iron abundance, and the metallicity of the model atmosphere was updated. This process was run twice with the updated metallicities to ensure convergence. (2) A new synthesis of all elements was generated

---

[4]MARCS model atmospheres at https://marcs.astro.uu.se/

[5]*linemake* contains laboratory atomic data (transition probabilities, hyperfine and isotopic substructures) published by the Wisconsin Atomic Physics and the Old Dominion Molecular Physics groups. These lists and accompanying line list assembly software have been developed by C. Sneden and are curated by V. Placco at https://github.com/vmplacco/linemake.

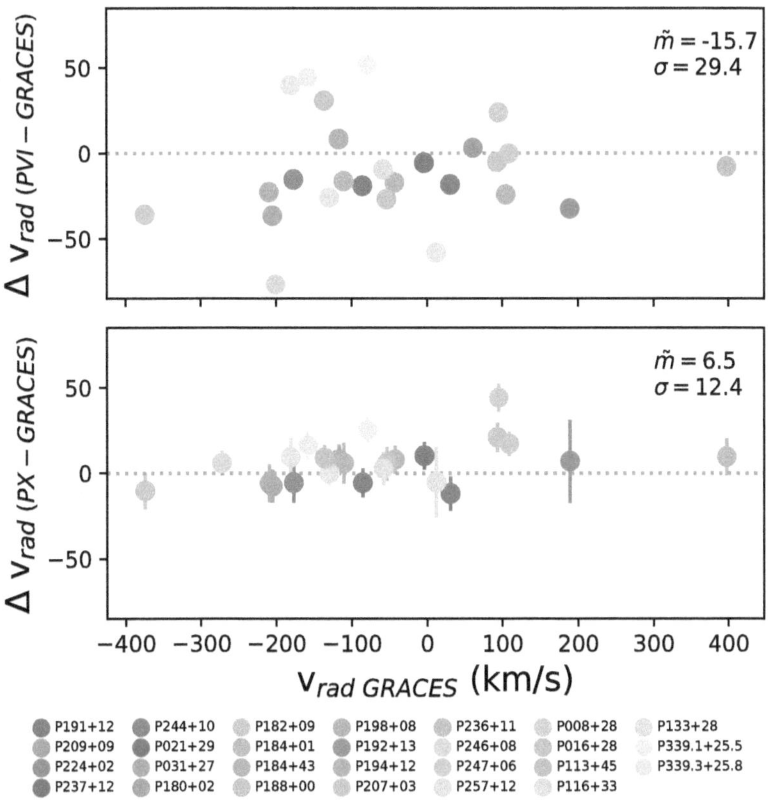

Figure 3.5 Comparison of radial velocities derived from *Pristine* medium-resolution spectroscopy and GRACES. MRS radial velocities in the top panel come from *Pristine* VI (PVI Aguado et al., 2019b) and from *Pristine* X (PX Sestito et al., 2020b) in the bottom panel. RV uncertainties were not reported with the Aguado et al. (2019b) velocities, but the error bars are included for the Sestito et al. (2020b) and GRACES data. The GRACES RV errorbars are smaller than the point size in this figure; thereby, systematic errors dominate the MRS RV values. Some GRACES targets are not shown as they do not have reliable velocities from the MRS. The median offset ($\tilde{m}$) and standard deviation (σ), both in km/s, are shown in each panel.

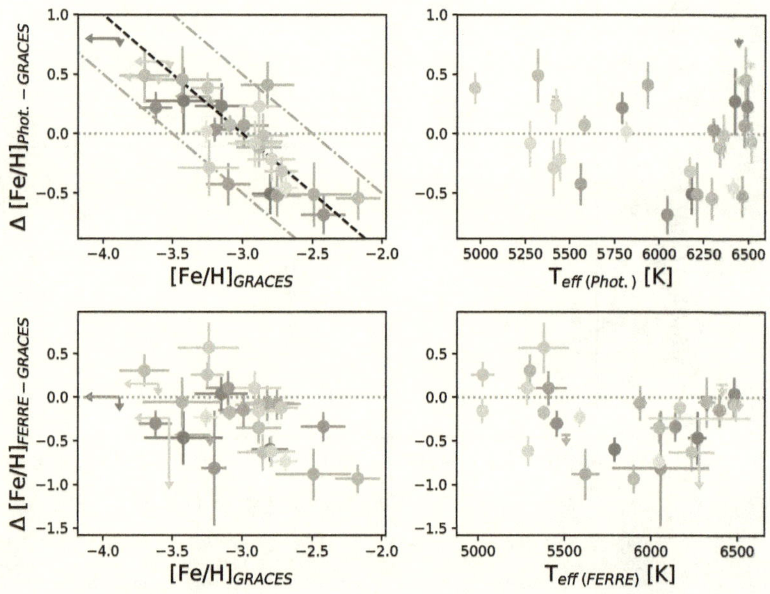

Figure 3.6 Comparison of [Fe/H] derived from photometry, MRS, and HRS with a Bayesian analysis utilizing Gaia parallaxes. Dashed lines of fixed [Fe/H] = −2.5, −3.0, and −3.5 are shown in the top left panel. Y-axes are shared between left and right panels.

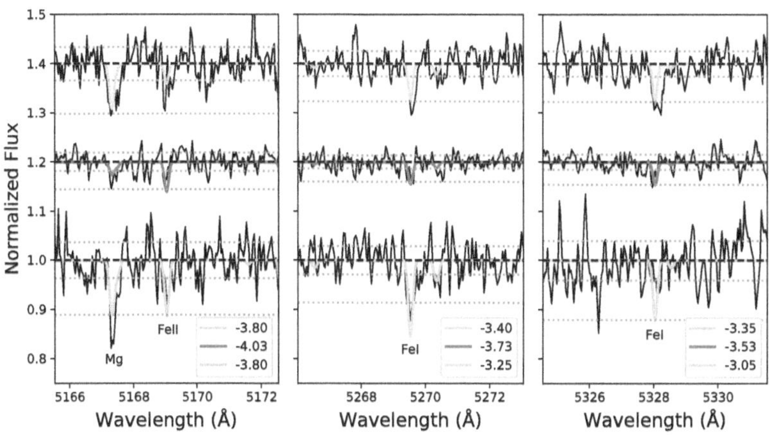

Figure 3.7 Synthesized Fe I and Fe II lines for the three stars with only Fe upper limits. Arranged top to bottom in order of hottest to coolest are P258+40 (T_{eff}=6686K), P237+12 (T_{eff}=6654K), and P246+08 (T_{eff}=6330K). Three Fe lines are shown (Fe II 5169.028Å, Fe I 5269.537Å, and Fe I 5328.039Å), chosen as they provide the tightest constraints on [Fe/H] for most stars in our sample. The GRACES spectra (solid black line) are compared to synthesized spectra (solid coloured lines that match the colour labels in Fig. 3.3 per star). The [Fe/H] abundances used in the synthetic spectra are noted in each panel. The continuum placement (dashed black line) and the $\pm1\sigma$ and -3σ noise levels (grey dotted lines) are shown.

including line abundances and upper limits for all of the clean spectral lines (e.g., see Figs. 3.7, 3.11, 3.10). (3) For the more challenging spectral features (i.e., those that were more severely blended or required hyperfine structure corrections), then a full spectrum synthesis was carried out using *linemake* to find all features and components within ±10 Å of the feature of interest. When a spectral feature was well fit, then the abundance was measured. If not, a 3σ upper limit was calculated. Each synthetic spectrum was broadened in MOOG using a Gaussian smoothing kernel with FWHM=0.15, other than two stars (P198+08 and P016+28) which required more broadening (FWHM=0.25). When there were multiple spectral lines for a given element, then the average of the measured abundances, or the lowest (most constrained) upper limit, was adopted. The number of lines synthesized per element ranged from 1 (Eu ɪɪ) to 25 (Cr ɪ). Blends, isotopic corrections, and hyperfine structure corrections were taken from *linemake* and included in the spectrum syntheses for lines of Li ɪ, O ɪ Sc ɪɪ, Mn ɪ, Cu ɪ, Zn ɪ, Y ɪɪ, Zr ɪɪ, Ba ɪɪ, La ɪɪ, and Eu ɪɪ. A sample line list is available in Appendix D; the full line list is available online.

Abundance errors are determined in two ways: (1) the line to line variations that represent measurement errors, e.g., in the continuum placement and/or due to the local SNR (see Tables D.11 to D.14), and (2) systematic uncertainties due to the stellar parameters (see Tables D.15 to D.20). The final abundance uncertainties are calculated by adding the line-to-line scatter (σEW) in quadrature with the uncertainties imposed by the stellar parameter errors (σT_{eff}, $\sigma\log g$, and σ[Fe/H]). These final uncertainties are used in the abundance plots (i.e., Figures 3.6, 3.8-3.9, B.4, B.2).

The EMP standard star HD122563 has been analysed from an archival very high SNR ($>$200) GRACES spectrum. We determined its stellar parameters using our Bayesian inference method, and compared those to literature values, in Venn et al. 2020. Those parameters and uncertainties have been adopted here. Our abundance results are similar to those in the literature and other Galactic standard stars (see Figs. 3.9, 3.12, and B.2.)

3.5.1 Iron-group

Iron abundances are calculated from the average of the synthetic fits to each spectral feature identified in Table D.23. Iron is calculated from Fe ɪ and Fe ɪɪ independently, where [Fe/H] in Table D.11 is the weighted average.

A comparison of the iron abundances between our high-resolution GRACES spectra, the MRS analysis, and the *Pristine* photometric estimates are shown in Fig. 3.6. In the top left panel, lines of constant metallicity are shown at [Fe/H] = -2.5, -3.0, and -3.5. This plot suggests that the precision in the current *Pristine* photometric metallicities may be limited to [Fe/H]$\geq$ -3, particularly for the hotter stars that dominate our sample. The *Pristine* team is working on complementary machine-learning methods to derive photometric metallicities, but it is yet unclear if these will result in better accuracy in the extremely metal-poor regime. In the top right panel, we find that our metallicities from the GRACES analyses suggest that the *Pristine* photometric metallicities near [Fe/H]$=-3$ have an intrinsic dispersion of [Fe/H]±0.5. This dispersion appears to be even larger for the FERRE results, as seen in the bottom left panel. The latter may also be temperature dependent, in that stars near $T_{\rm eff}$ (FERRE) = 6000 K result in [Fe/H]$_{\rm MRS}$ (FERRE) metallicities that are too low, as seen in the bottom right panel.

Samples of the spectrum synthesis of three Fe lines (and one Mg line) are shown in Fig. 3.7 for three stars: P258+40 ($T_{\rm eff}$=6686 K), P237+12 ($T_{\rm eff}$=6645 K), and P246+08 ($T_{\rm eff}$= 6330 K). P237+12 is the lowest metallicity star in our sample. We determine an upper limit of [Fe/H]$\leq$ -3.9, despite a very high SNR spectrum (SNR$\sim$95 at 6000 Å); this is partially due to the limited (red) wavelength region available in the GRACES spectra and lack of strong Fe II lines in this wavelength region. Examination of all three spectral lines shown in Fig. 3.7 provide a mean 1DLTE 3σ upper limit for P237+12 of [Fe/H]$=<$ -3.9. A follow-up VLT-UVES spectrum of this star confirms this low metallicity from more and bluer iron features (Lardo et al., in prep.).

Departures from LTE are known to overionize the Fe I atoms due to the impact of the stellar radiation field on the level populations, particularly in hotter stars and metal-poor giants. These non-LTE (NLTE) effects can be significant in our stellar parameter range. To investigate the impact of NLTE corrections on the iron abundances, the [Fe I/H]$-$[Fe II/H] differences are compared to the surface gravities in Fig. 3.8. We find that [Fe II/H] is usually lower than [Fe I/H]$_{\rm 1DLTE}$ (top panel) by $\sim$ 0.15 dex, which is *not* the typical signature of neglected NLTE effects. NLTE corrections are examined from two databases; (1) INSPECT[6] provides NLTE corrections for some of our Fe I and Fe II lines (Bergemann et al., 2012; Lind et al., 2012), and (2) the MPIA NLTE grid[7] which provides NLTE correction for several additional lines (Bergemann et al., 2012; Kovalev et al., 2019). The NLTE corrections for our Fe II lines are negligible. When the Fe I NLTE corrections are applied (bottom panel), the difference increases such that $<$[Fe I/H]$_{\rm NLTE}-$[Fe II/H]$>\sim$ +0.35. The average NLTE corrections per star are summarized in Table D.11. We attribute the differences in [Fe I/H]$_{\rm NLTE}-$[Fe II/H] vs. surface gravities in Fig. 3.8 to the potential systematic errors

---

[6]INSPECT NLTE corrections are available at http://inspect-stars.net.

[7]MPIA NLTE corrections are available at http://nlte.mpia.de.

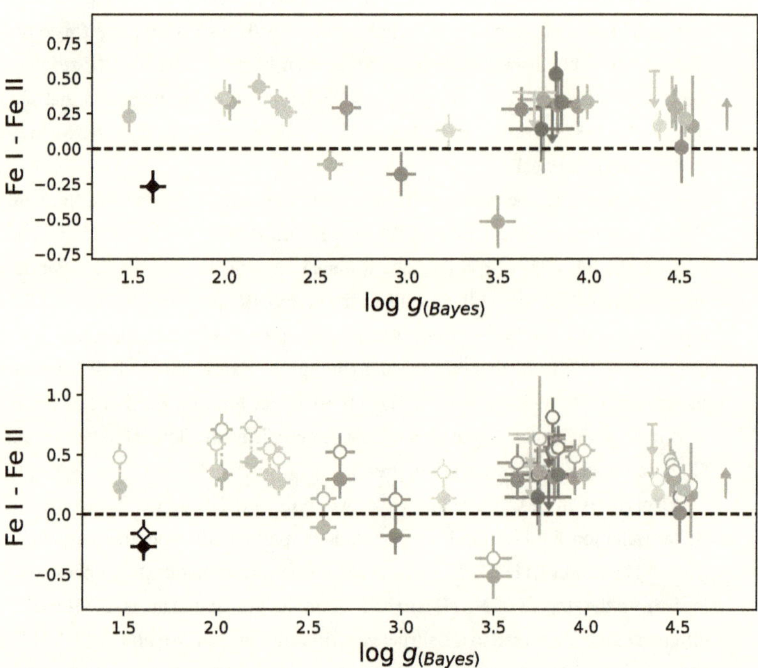

Figure 3.8 Comparison of [Fe I/H] - [Fe II/H] vs. surface gravity derived from our Bayesian inference method. Upper panel are 1DLTE abundances, and lower panel includes NLTE corrections (open circles; see text). [Fe II/H] is lower than [Fe I/H] for a majority of the stars. The dashed black line at [Fe I/H] - [Fe II/H] = 0 represents ionization equilibrium - a spectroscopic check for the accuracy of the surface gravity. Reference star HD 122563 is shown as the black diamond with log g = 1.61 ± 0.1 from Venn et al. (2020).

in the metal-poor MIST/MESA isochrones adopted in the Bayesian inference method (and discussed previously in Section 4). Considering Fe II lines are much more gravity sensitive than Fe I lines, blue sensitive spectra is needed to further constrain the Fe II abundances and surface gravities.

Three stars have lower Fe I than Fe II abundances, P016+28, P224+02, and P198+08. We notice that the first two are flagged as carbon-enhanced red giants. This may have affected our Bayesian inferences using MIST/MESA isochrones that are based on carbon-normal stars. We do not explore this further as more detailed analysis of the impact of carbon on synthetic stellar spectra and the MIST/MESA isochrones is beyond the scope of this study.

The other iron-group elements Cr and Ni were also synthesized, but only 1-4 spectral lines of each element are available in our GRACES spectra. Their abundances are provided in Table D.13 and shown in Fig. 3.9. When Cr or Ni are calculated, their 1DLTE abundances are in excellent agreement with Fe, which is similar to other metal-poor stars in the Galaxy. Examination of the NLTE corrections suggests that [Cr/Fe] may be increase by up to ~ 0.4 This would be important for the detailed interpretation of the chemistry of each star; however, we do not include the NLTE effects in Fig. 3.9 since the Galactic comparison stars are not also corrected.

3.5.2 Carbon

As the GRACES spectra are restricted to redder wavelengths, we do not have any carbon features to analyse. However, [C/Fe] is determined in the FERRE analysis of our MRS from the G-band and features below 4300 Å. In Table D.9, we identify our targets that were reported to have C-enhancements by Aguado et al. (2019b). A total of eight of the 30 stars in our sample were identified by FERRE to have [C/Fe] > +1.0 (P016+28, P021+29, P113+45, P133+28, P184+01, P188+00, P224+02, P339.1+25.5). Unfortunately, the FERRE pipeline struggles to determine C when the SNR of the MRS is low ($\lesssim 25$) in the temperature range of our targets. Closer inspection of the MRS shows that five of these are likely upper limits to non-existent G-bands.

Only three stars appear to have noticeable G bands in the MRS spectra, and in each of these cases the FERRE analysis found them to be very carbon-rich, with [C/Fe]> +2 (P016+28, P184+01, P224+02). Only P016+28 is flagged to have a reliable carbon abundance. The other two do appear to be C-enhanced, but their specific [C/Fe] values are quite uncertain. The slight changes in our stellar parameters from the Bayesian inference method are also likely to affect the final C abundances.

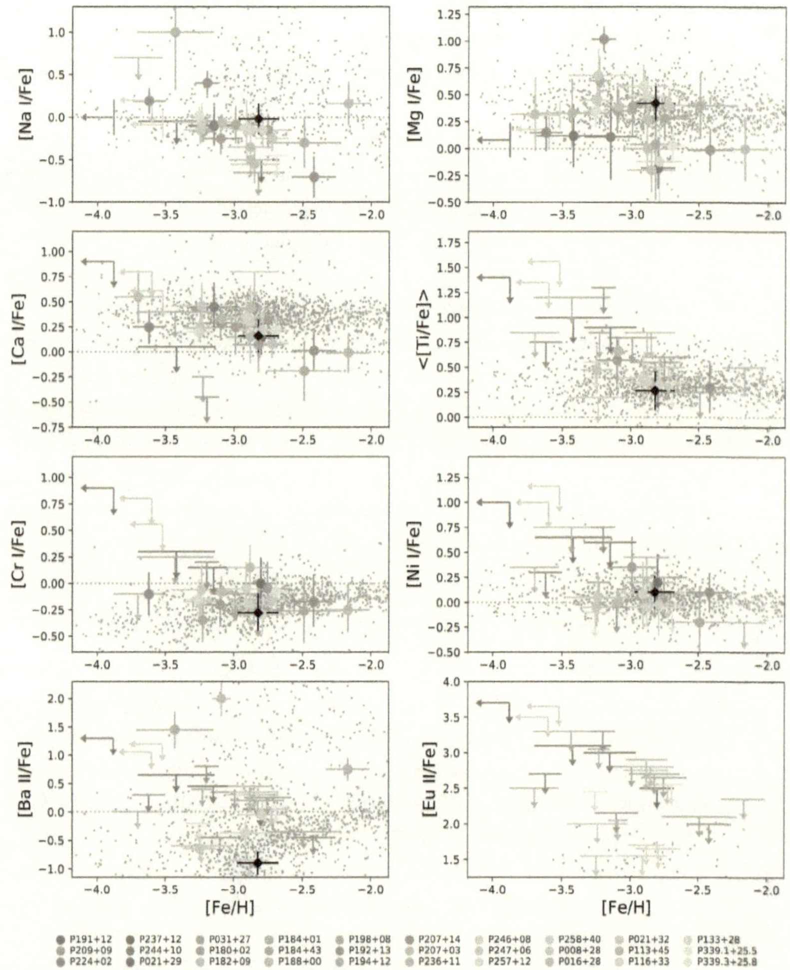

Figure 3.9 Elements with good spectral line detections, and Eu II . <[Ti/Fe]> is the average of [Ti I/Fe] and [Ti II/Fe] when a detection was made for both species, otherwise it is the highest upper limit between the two species. Our analysis of the GRACES spectrum for HD 122563 is shown as the black diamond. The Galactic reference star abundances are taken from the literature (Venn et al., 2004; Aoki et al., 2013; Yong et al., 2013; Roederer et al., 2014; Frebel et al., 2014).

In summary, we find the C abundances for most of our targets are not sufficiently reliable from the MRS analysis; nevertheless, one star is clearly C-rich (P016+28), and two others are very likely C-enhanced (P184+01, P224+02).

3.5.3 Alpha elements

Alpha elements in extremely metal-poor stars form primarily though He-capture during various stages in the evolution of massive stars, and dispersed through SNe II events. Thus, the [α/Fe] ratio is a key tracer of the relative contributions of SN II to SN Ia products in a star forming region. In this study, the α-elements include Na, Mg, Ca, and Ti. We have included Na because it typically scales linearly with Mg in metal-poor stars in the Galaxy (e.g., Pilachowski et al., 1996; Venn et al., 2004; Norris et al., 2013). We have also included Ti since it too scales with other α-elements even though the dominant isotope ^{48}Ti forms primarily through Si-burning in massive stars (e.g., Woosley et al., 2002).

Na abundances or upper limits are from the two strong Na D lines. These lines are clear and present in most of the stars in our sample, and easily separated from any interstellar lines. NLTE effects can be significant for these resonance lines, however corrections in this analysis are small (ranging from Δ(Na) = -0.1 to -0.2 for most stars), as shown in Table D.22. Mg is from 2-4 lines of Mg I in all stars, even P237+12 which has only an iron upper-limit. NLTE corrections are small (typically ΔMg$\leq$+0.2). Ca is from 1-10 lines of Ca I in most stars, or the Ca I 6122.217 and Ca I 6162.173 lines are used to estimate an upper limit (e.g., see Fig. 3.11). NLTE corrections are moderate (typically ΔCa$\leq$+0.3). The Ca II triplet is also examined, however we do not use those results in our analysis (especially without NLTE corrections). Ti is from 1-4 lines of Ti I and 1-6 lines of Ti II. For many stars, upper limits only were available and estimated from Ti I lines. The average NLTE corrections for Ti I can be large (ΔTi $\leq$+0.6), however the NLTE corrections to Ti II are negligible. Again, we do not include the NLTE corrections in Fig. 3.9 since abundances for the Galactic comparison stars are not also corrected.

The majority of these newly discovered extremely metal-poor stars show 1DLTE abundances that are within 1σ uncertainties of the Galactic comparison stars, especially given our results for the EMP standard star, HD 122563 (black point in each abundance plot). Only a handful of stars have α-element abundances that are statistically lower than the Galactic comparison stars. These include;

1. Three stars with [Fe/H]$\geq$ -2.5 (P113+45, P192+13, P198+08), in particular they all show scaled-solar ratios of [Ca/Fe], rather than the higher plateau value near

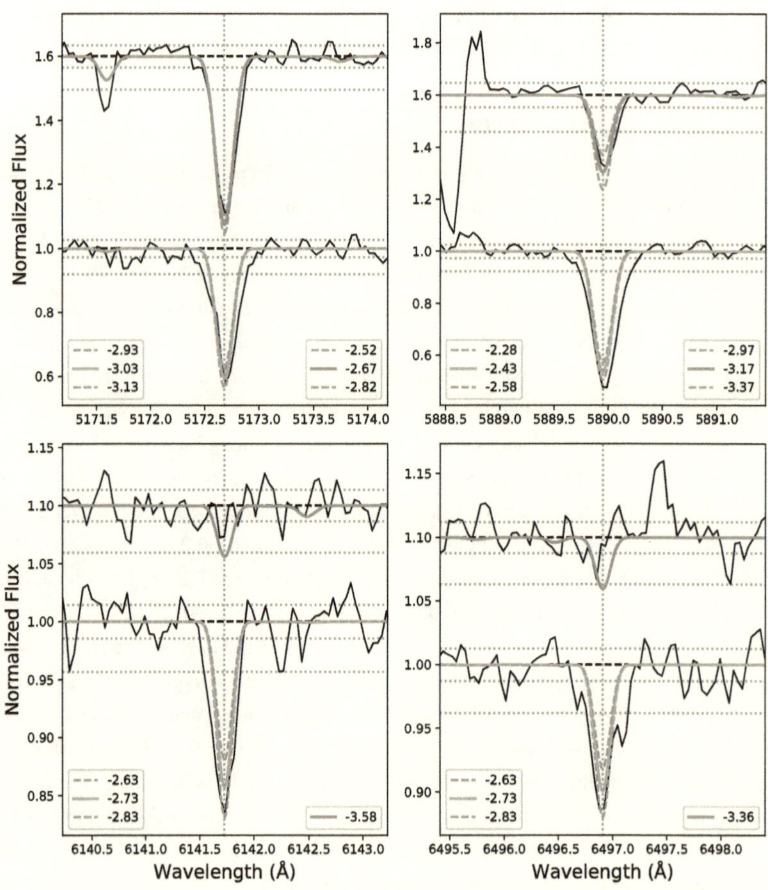

Figure 3.10 Syntheses of Mg I 5172.68Å and Na I 5889.95Å (top panels), and two Ba II lines (6141.73Å, 6496.91Å; bottom panels) for P184+01 (blue, bottom) and P192+13 (red, top). These lines provide abundances or the tightest constraints for these two stars. The GRACES spectrum is the solid black line, and the synthesized spectra are the solid colored lines with the [X/H] measurements in the legend. Dashed grey lines represent the $\pm 1\sigma$ synthesis. The dashed black line represents the continuum placement and the dotted grey lines are $\pm 1\sigma$ and -3σ, where σ is defined as the measured scatter in the continuum.

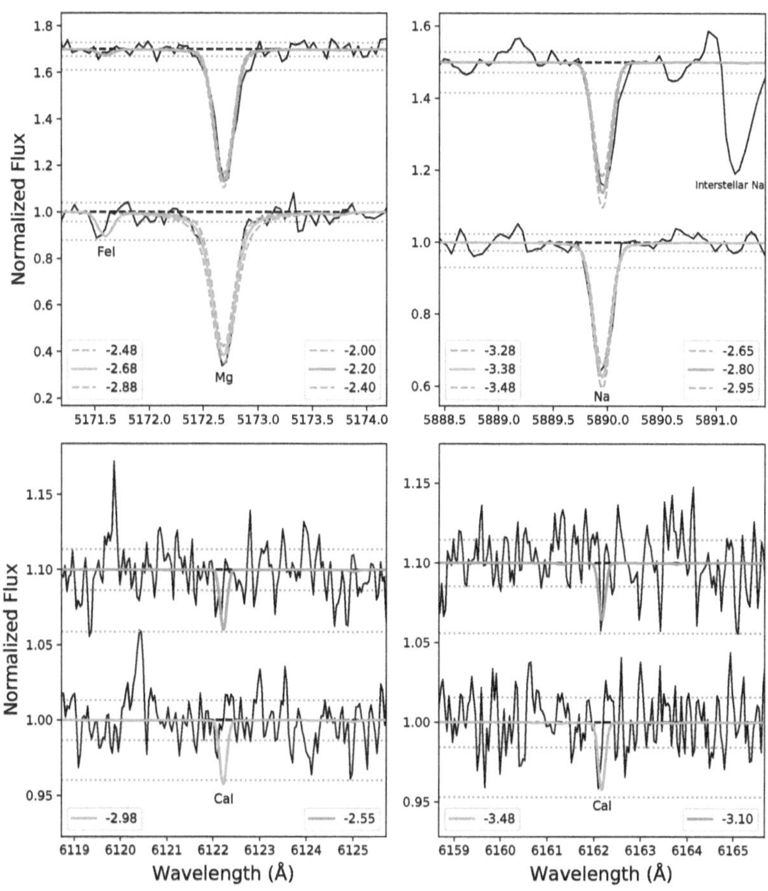

Figure 3.11 Syntheses of Mg I 5172.68Å and Na I 5889.95Å (top panels), and two Ca I lines (6122.21Å, 6162.17Å; bottom panels) for P207+14 (brown, top) and P184+43 (blue, bottom). These lines provide abundances or the tightest constraints for these two stars. Both stars have [Fe/H]= −3, and are notably Mg-rich, but Ca-poor. See Fig. 3.10 for additional label information.

[Ca/Fe]=+0.4. This Ca-depletion is supported by low [Mg/Fe] and/or low [Na/Fe] abundances, and NLTE corrections only lower these element ratios further, indicative of origins in accreted dwarf galaxies. To demonstrate that these low abundances are real, we show the observed and synthetic spectra around the Mg I 5172.68 and Na I 5889.95 lines in P192+13 in Fig. 3.10.

2. Three other stars near [Fe/H]=−3 show sub-solar [Mg/Fe] values, yet normal plateau values of [Ca/Fe] (P021+29, P133+28, P182+09). This pattern has been seen for stars in dwarf galaxies, and is typically attributed to an effectively truncated upper IMF, loss of gas from high mass supernova events, and/or inhomogeneous mixing of supernova yields in the dwarf galaxy's interstellar medium (e.g., Tolstoy et al., 2009; McWilliam et al., 2013; Kobayashi et al., 2015; Frebel & Norris, 2015).

3. One star near [Fe/H]= −3.5, which shows an enrichment in [Na/Fe] (P184.1783+01.0664). This star will be discussed further below (see Section 3.6.1).

4. Two stars with [Fe/H]< −3 have a very rare abundance signature. P184.2997+43.4721 and P207.3454+14.1208 show low, sub-solar, upper limits in [Ca/Fe], which is unusual when compared with the Galactic halo sample, especially as they are both enriched in [Mg/Fe] (> +0.5). Their spectra and our 1DLTE synthesis of a sample of their Mg I, Na I, and Ca I lines are shown in Fig. 3.11. The NLTE corrections are small and do not affect these trends (see Table D.22). This abundance pattern has been seen in only a few EMP stars (see Sitnova et al., 2019), and is discussed further below.

3.5.4 Neutron-capture elements

Elemental abundances or upper limits are determined for six neutron capture elements: Y, Zr, La, Nd, Ba, and Eu. These formed in massive stars and core collapse supernovae through rapid neutron capture reactions, and those other than Eu also form via slow neutron captures during the thermal pulsing AGB phase in intermediate-mass stars. The specific details and yields from these nucleosynthetic processes is a dynamic field of current research. For the core collapse SNe, new models and calculations of their yields include details of the SN explosion energies, explosion symmetries, early rotation rates, and metallicity distributions (e.g., Kratz et al., 2007; Nishimura et al., 2015; Tsujimoto & Nishimura, 2015; Kobayashi et al., 2020), as well as exploration of contributions from compact binary mergers as a (or as the most) significant site for the r-process (e.g., Fryer et al., 2012; Korobkin et al.,

2012; Côté et al., 2016; Emerick et al., 2018). Similarly, predictive yields from AGB stars by mass, age, metallicity distributions, and details of convective-reactive mixing are also an active field of research (e.g., Lugaro et al., 2012; Cristallo et al., 2015; Pignatari et al., 2016).

Only Ba and Eu are discussed in this section. For all stars, we have 1-2 Ba II lines (6141.73 Å, 6496.91 Å). Hyperfine structure and isotopic splitting are taken into account using the atomic data in *linemake*. For Eu, the GRACES spectra only permit studies of the Eu II 6645 Å line, which is too weak to be observed in any of our spectra. The upper limits for Eu from this line are also too high to be scientifically useful in testing for pure r-process enrichment in these stars. It would be important to examine the much stronger Eu II 4129 Å line to constrain the pure r-process contributions in these stars.

Two stars appear to be Ba-rich (P184+01 and P016+28); as seen in Fig. 3.9, both appear to have [Ba/Fe]~ +1 with [Fe/H]< −3. The spectrum of these Ba II lines in P184+01 is shown in Fig. 3.10, where it is clear that these lines are clear, strong, and well measured. NLTE corrections were calculated for these two Ba-rich stars following the methods in Mashonkina et al. (1999) and Mashonkina et al. (2019), using models representing their specific stellar atmospheres and high LTE Ba abundances. The NLTE corrections for P184+01 increase [Ba/Fe] by 0.09 dex, when averaged between the individual corrections for the 6141Å and 6496Å lines. Alternatively, [Ba/Fe] decreases by 0.44 dex when NLTE corrections are calculated for P016+28. Including these corrections, the NLTE Ba abundances are still significantly higher than the 1DLTE results in similar EMP stars in the Galaxy.

One star appears to be Ba-poor (P192+13). In Fig. 3.10, the spectrum for P192+13 around the Ba II 6141 and 6497Å lines shows that these lines are not present. Using the 3-σ line depth, we find a very low upper limit abundance, [Ba/Fe] < −1 at [Fe/H]=−2.4 (see Fig. 3.9). Such a low Ba abundance at this metallicity is unusual for stars in the Galaxy, but more common in ultra faint dwarf galaxies, such as Segue I, Com Ber, and Hercules (e.g., Koch et al., 2013; Frebel et al., 2014; Ji et al., 2016a).

The other neutron capture elements with spectral features in the GRACES wavelength regions are examined in Appendix B.1.2. Only upper limits could be determined, and they did not provide useful scientific constraints.

Lithium

The Li I 6707 spectral line is present in the spectra of over half of our targets. Our spectrum syntheses for Li included hyperfine structure and isotopic splitting, with atomic data taken

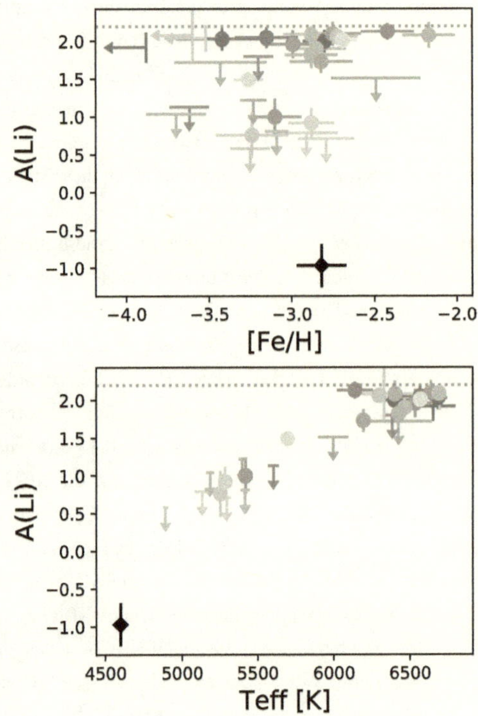

Figure 3.12 Lithium abundances from Li I 6707 Å. Stellar labels are the same as in Fig. 3.9. The Spite plateau (Spite & Spite, 1982; Sbordone et al., 2010) is shown as the dashed gray line.

from *linemake*. The study of lithium in EMP stars is an active topic of discussion due to the links between EMP stars and the chemistry of the early Universe. The cosmological lithium problem refers to the discrepancy between the amount of Li predicted from Big Bang nucleosynthesis (A(Li) = 2.67 - 2.74; Cyburt et al., 2016; Coc & Vangioni, 2017) and the highest Li abundances measured in the atmospheres of unevolved metal-poor stars (A(Li)~2.2, the Spite Plateau; Spite & Spite, 1982; Bonifacio et al., 2007; González Hernández et al., 2008; Aoki et al., 2009; Sbordone et al., 2010). Finding unevolved EMP stars with detectable Li provides strong constraints on the lithium problem.

A majority of our sample shows expected trends between A(Li), T_{eff}, and metallicity (Fig. 3.12). Most of the hotter and higher metallicity stars are found near the Spite plateau. As metallicity decreases, a higher degree of scatter in A(Li) is observed, consistent with the meltdown of the Spite plateau observed by Sbordone et al. (2010); Bonifacio et al. (2012). The RGB stars in our sample show lower lithium abundances (or upper limits), as expected since Li is destroyed through surface convection in cooler stars (Spite & Spite, 1982). Three stars, however, are notable: P237+12, P246+08, P258+40. All three stars only have metallicity upper-limits with [Fe/H]< -3.5, but have detectable Li at the 3σ level. Their measured A(Li) places them at the Spite plateau (see Table D.15). Similar in T_{eff}, $\log g$, metallicity, and A(Li) to the primary star of the spectroscopic binary CS22876-032 (González Hernández et al., 2008, 2019), these main-sequence and turn-off stars are excellent candidates for follow-up studies related to the cosmological lithium problem.

3.6 Discussion

Using the chemical abundances determined in this chapter, we discuss our EMP stars in terms of the accretion history and chemical evolution of the Galaxy.

3.6.1 New CEMP candidates

In this chapter, we have analysed three stars with slight to large carbon enhancements. As their precise [C/Fe] results are quite uncertain, we regard these stars simply as carbon-enhanced metal-poor (CEMP) candidates, rather than confirmed CEMP stars. At the lowest metallicities, stars are often found to be enhanced in carbon (Beers et al., 1992; Norris et al., 1997), typically comprising 40% of the EMP stars (see Yong et al., 2013; Lee et al., 2013; Placco et al., 2014), though recently those percentages have been lowered through considerations of the carbon 3DNLTE corrections (Norris & Yong, 2019).

Different types of CEMP stars have been identified and defined by Beers & Christlieb (2005), where the two main classes are the CEMP-s stars, which show additional enhancement in s-process elements (such that [C/Fe] > +0.7 and [Ba/Fe] > +1.0), and the CEMP-no stars, which do not show any s-process enhancements. These initial definitions have been further refined by Yoon et al. (2016), based on the trends observed between [Fe/H] and [C/Fe]. The C excess in CEMP-no stars is generally attributed to nucleosynthetic pathways associated with the very first stars to be born in the universe (Iwamoto et al., 2005; Meynet et al., 2006).

Of our three new CEMP candidates, we find that two (P016+28 and P184+01) are enriched in barium, with [Ba/Fe]> +1 (to within their 1σ uncertainties), which suggests that they may belong to the CEMP-s sub-class. Examining P016+28 further, the carbon abundance from FERRE is [C/Fe]=+2.42, for an absolute carbon abundance of A(C)=7.76. If this carbon abundance is accurate, this star would be amongst the high-C/Group I population in Yoon et al. (2016, A(C) = 7.96 ± 0.42). Meanwhile, P184+01 has [Fe/H]=-3.43 and a very uncertain carbon abundance of A(C)=7.26, which places it between Groups I, II, and III. Another way to test the CEMP-s hypothesis for these two stars is through radial velocity monitoring, as most CEMP-s stars are found in a binary systems (e.g., Lucatello et al., 2005; Hansen et al., 2016; Starkenburg et al., 2014, and references therein). This property has contributed to the theory that CEMP-s stars have received their carbon and s-process enhancements through mass-transfer with an asymptotic giant branch (AGB) star in a binary system (Abate et al., 2013). We do not search for radial velocity variations in our data though, since the GRACES spectra were rarely taken over several epochs, and the medium-resolution INT spectra do not have sufficient precision.

The remaining CEMP-no candidate, P224+02, has a low barium upper-limit. As this star has [Fe/H]= −3.62 (and a tentative A(C) = 7.29), it falls between the A(C)-metallicity groups in Yoon et al. (2016). Groups II and III are dominated by CEMP-no stars, but recent analysis of Group I CEMP stars has shown that 14% are CEMP-no (Norris & Yong, 2019). While generally low in s-process elements, the Group I CEMP-no stars also show higher [Sr/Ba] abundances than the majority of Group I CEMP-s/rs stars. To explain the [Sr/Ba] ratios, Norris & Yong (2019) speculated that Group I CEMP-no stars may experience some mass exchange with massive AGB stars or rapidly-rotating "spinstars" in binary systems (both produce less s-process material). Unfortunately, we do not determine [Sr/Fe] as the Sr II lines are at blue wavelengths, not reached by our GRACES spectra. Alternatively, we examine the α-elements (Na, Mg) since some CEMP-no stars can show enrichments in these elements; however, as discussed by Maeder & Meynet (2015), the predictions for these elements depend on mixing in massive stars and the predictions can vary widely. Furthermore, Frebel & Norris (2015) show that enhanced alpha-elements only occur in about half of their CEMP-no sample, and even less when [Fe/H]$\sim$ -3. The exact origin(s) of the CEMP-no stars is not yet clear, however it has also been proposed that CEMP-no stars may form in dwarf galaxies, and thereby may be associated with accreted systems (Yuan et al., 2020; Limberg et al., 2020). To examine this further, we compute the orbits of our 30 GRACES stars below.

3.6.2 Other stars with interesting chemistries

Highlighted in Section 3.5.3, P184+43 and P207+14 both show very low, sub-solar, upper limits in [Ca/Fe] (< -0.3) and large [Mg/Fe] enrichement ($> +0.5$). This results in [Mg/Ca]$\sim$+0.8. This rare abundance pattern had been seen in only a few EMP stars (see Sitnova et al., 2019). One star that was highlighted by Cohen et al. (2007), HE1424-0241, which has a similarly high [Mg/Ca] = +0.83, mostly driven by deficient Ca, but at a slightly lower metallicity [Fe/H]= -4. This abundance pattern is interpreted as contributions to stellar Mg and Ca abundances from only a small number of SN II explosions, i.e., where the nucleosynthetic yield for explosive alpha-burning nuclei like Ca was very low compared to that for the hydrostatic alpha-burning element Mg. Similarly, two stars in the Hercules dwarf galaxy studied by Koch et al. (2008) display [Mg/Ca] = +0.58 and +0.94 dex. Koch et al. (2008) argues that these high ratios can be attributed to enrichment from high mass ($\sim 35 M_\odot$) Type II SNe, based on yields from Woosley & Weaver (1995). Furthermore, their chemical evolution models for Hercules-like dwarf galaxies indicate that the observed [Mg/Ca] ratios can only be reproduced in 10% of the systems that are enriched by only a few (1-3) Type II events. Clearly these are chemically unique objects which reflect the chemical evolution of rare environments. Sitnova et al. (2019) also note that these exceptional stars only comprise $<10\%$ of stars with [Fe/H]< -3 and do not typically reveal carbon enhancement.

3.6.3 Orbit calculations

Gaia DR2 proper motions and astrometry have dramatically accelerated the fields of Galactic Archaeology and near-field cosmology by providing the data needed to calculate the detailed orbits of nearby stars. The stars in this study were also selected to have small parallax errors, and therefore precise distances. When combined with the precision radial velocities from

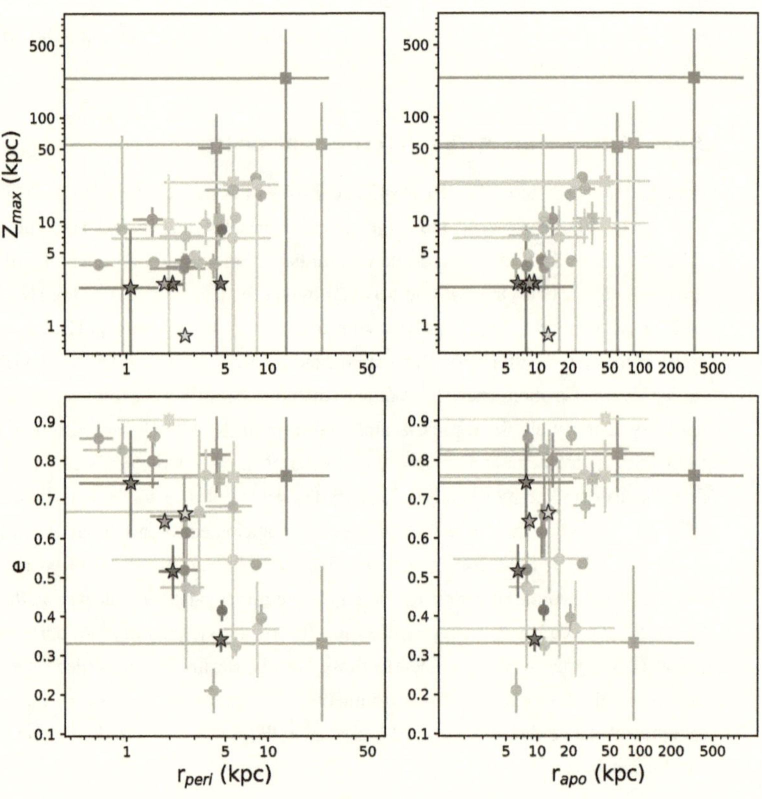

Figure 3.13 Orbital elements for stars from 2018A-2019B datasets. Star symbols represent stars in planar disk orbits, squares are stars that reach the outer Galactic halo.

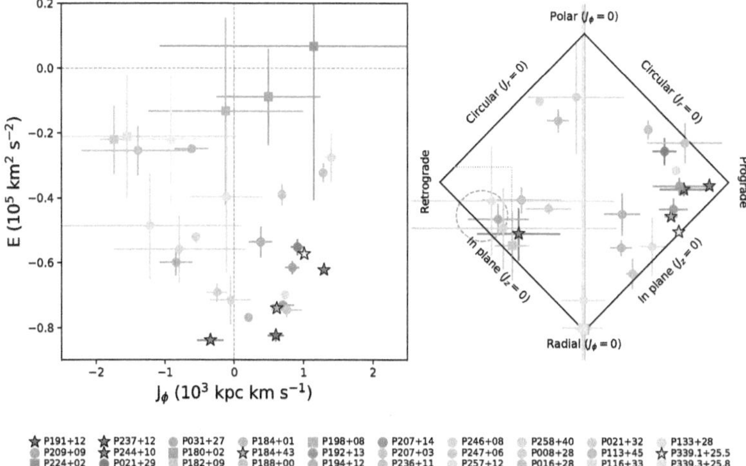

Figure 3.14 Action vectors and energies for stars from 2018A-2019B datasets. Same symbols as in Fig. 3.13. The dotted grey box in the right panel is the dynamical cut for Gaia-Sequoia from Myeong et al. (2019) ($e \sim 0.6$ with $J_\phi/J_{tot} < -0.5$ and $(J_z - J_R)/J_{tot} < 0.1$) and the dashed grey ellipse represents the mean dynamical properties of the associated Gaia-Seqouia stars from Limberg et al. (2020).

our high-resolution GRACES spectra, then we are able to estimate the orbits for these stars to within the accuracy of our assumptions on the MW potential.

Orbital parameters for the stars in this study are calculated with *Galpy* (Bovy, 2015), using the parallax distances, our radial velocities from the GRACES spectra, and the *Gaia* DR2 proper motions. The *MWPotential14*[8] was adopted, though a more massive halo of $1.2x10^{12}M\odot$ was chosen following Sestito et al. (2019). Errors have been propagated from the uncertainties in the proper motions, RVs, and distances via Monte-Carlo sampling of the Gaussian distributions of the input quantities.

The apocentric and pericentric distances (R_{apo} and R_{peri}), perpendicular distance from the Galactic plane (Z_{max}), and eccentricity (e) of the calculated orbits for our stars are shown in Fig. 3.13. Following Sestito et al. (2019), stars with $R_{apo}<15$ kpc and $|Z|_{max}<3$ kpc are considered to be confined to the Galactic plane, while stars with $R_{apo}>30$ kpc are considered to be members of the outer halo. The orbital energy (E) and action parameters (vertical J_z, azimuthal J_ϕ) for the sample are also calculated with *Galpy*, and shown in Fig. 3.14. All targets appear to be bound to the Milky Way, to within their uncertainties.

3.6.4   Stars with Interesting Orbits

Five of our new EMP stars (P184+43, P191+12, P237+12, P244+10, and P339.1+25.5) have distinctly planar-like orbits ($|Z|_{max} < 3$ kpc). All five have [Fe/H]< -3 and somewhat elliptical orbits (e=0.3 to 0.7). P237+12 is the most metal-poor star in the GRACES sample with a 1DLTE metallicity upper-limit of [Fe/H]< -3.88. This star has Mg below the canonical MW halo plateau values, though the abundances, when available, are generally within the regime of normal Galactic halo stars. The implications of finding planar-like EMP stars, in the context of Galactic evolution, is discussed further in the following section. P339.1+25.5 stands out dynamically as it has the lowest $|Z|_{max}$ (< 1 kpc) of the stars in this sample, suggesting it may be coincident with the prograde Galactic thin disk. However, its somewhat elliptical orbit ($e \sim 0.7$) is uncharacteristic of typical thin disk stars.

The star P184+43 with a prograde, planar orbit also has an unusual chemistry, as it is enriched in magnesium, [Mg/Fe]=+0.6, and yet has a very low upper-limit on calcium, [Ca/Fe]< -0.2. In Section 3.6.2, we speculated that this peculiar abundance pattern may suggest a lack of contributions from lower mass stars and SN Ia. If so, then this star could have formed in a dwarf galaxy that was accreted by the Milky Way at early times, before

---

[8]This potential is three component model composed of a power law, exponentially cut-off bulge, Miyamoto Nagai Potential disc, and Navarro, Frenk & White (1997) dark matter halo.

SN Ia could enrich it. This speculation is discussed further below (Section 3.6.5).

Five high-energy, highly-retrograde stars (P016+28, P021+32, P031+27, P133+28, P182+09) appear to be dynamically related to the "Gaia-Sequoia" accretion event (Myeong et al., 2019). Sequoia is the population of high-energy retrograde halo stars that are presumed to be associated with an ancient accretion event of a counter-rotating progenitor dwarf galaxy (Myeong et al., 2019; Matsuno et al., 2019; Monty et al., 2020; Cordoni et al., 2020; Yuan et al., 2020; Limberg et al., 2020). Myeong et al. (2019) identifies stars with $e \sim 0.6$, $J_\phi/J_{tot} < -0.5$, and $(J_z - J_R)/J_{tot} < 0.1$ to be linked to Gaia-Sequoia. These five stars also meet the more recent and stricter membership criteria by Limberg et al. (2020, see Fig. 3.14), to within their errors. We acknowledge that these simple action vector cuts do not account for background stars which are non-coincidental with Sequoia. Limberg et al. (2020) explored the effect of background contamination via a membership clustering algorithm and found that up to $\sim 50\%$ of their Sequoia sample may indeed be contamination. Due to the small size of our sample, we do not explore this avenue further and only offer these stars as Gaia-Sequoia candidates to be confirmed/rejected in later studies. These add to the few extremely metal-poor stars now found associated with Gaia-Sequoia (see also Monty et al., 2020; Cordoni et al., 2020; Limberg et al., 2020).

The action vectors computed for the orbit of P244+10 suggests that it may also be associated with Gaia-Sequoia; however, its low energy planar-like orbit is uncharacteristic of other Sequoia members. P244+10 is more dynamically similar to stars associated with the Thamnos event (Helmi et al., 2017; Koppelman et al., 2018; Limberg et al., 2020). Thamnos is also believed to be a lower metallicity - higher α-abundance population than Gaia-Seqouia (Koppelman et al., 2018; Monty et al., 2020; Limberg et al., 2020), however our chemical abundances are in good agreement with our other Gaia-Sequoia candidates, or possibly even lower in the hydrostatic α-elements (like [Mg/Fe]). Regardless of its detailed chemistry, P244+10 remains a kinematically interesting star for follow-up investigations.

We note that the parallax distances are slightly different from the distances inferred from our Bayesian inference method (discussed above). In particular, distances are affected in the Bayesian inference analysis through the application of MIST/MESA isochrones. This is discussed further in Appendix B.2. We suggest that the parallax distances are more reliable, and provide the most accurate orbit estimates to within the accuracy of the Milky Way Galaxy potential that we have adopted.

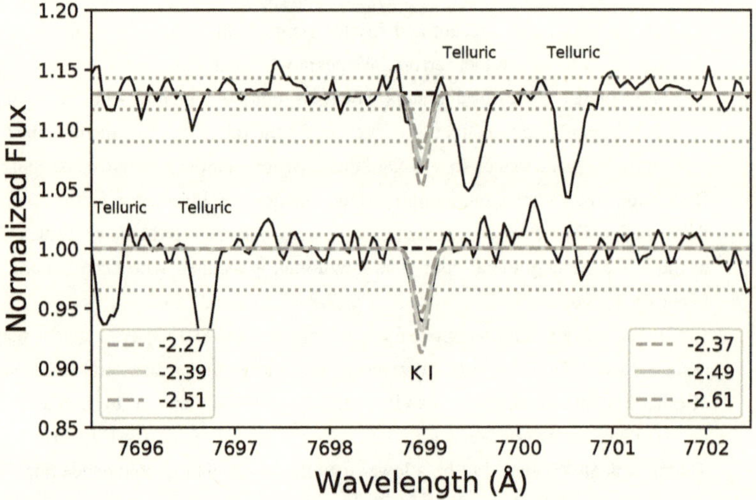

Figure 3.15 Syntheses of K I at 7698.97 Å. P133+28 (light grey, bottom) and P021+32 (light purple, top). Both objects are K-rich (e.g. Fig. B.2) and appear to be dynamically related to the Gaia-Sequoia accretion event (e.g. Fig. 3.14). Telluric absorption lines are identified, however telluric subtraction was not performed as no spectral standards were observed in tandem with these science acquisitions. See Fig. 3.10 for additional label information.

3.6.5 Formation and Early Chemical Evolution of the Galaxy

This work is consistent with the results by Sestito et al. (2019) who found that a large number of ultra metal-poor stars in the literature are on planar orbits in the Galaxy. We have found five new EMP stars on planar-like orbits (or 16% of our sample). Four are on prograde orbits and one (P244+10) is on a retrograde orbit. These ratios are comparable to our earlier results from an analysis of CFHT spectra for 115 metal-poor candidates which resulted in 16 very metal-poor stars ([Fe/H]< −2.5) confined to the Galactic plane (R_{apo} < 15 kpc, Z_{max} < 3 kpc), including one on an extremely retrograde orbit (Venn et al., 2020). In comparison with the larger *Pristine* database of medium-resolution INT spectra, Sestito et al. (2020a) found that the ratio of metal-poor stars ([Fe/H]< −2.5) with prograde versus retrograde orbits is 1.28, based on 358 stars. We find a similar ratio of 1.23, for our sample of 29 stars with [Fe/H]< −2.5. When compared to the Numerical Investigation of a Hundred Astronomical Objects (NIHAO Wang et al., 2015) hydrodynamic simulations of galaxy formation, specifically the NIHAO-UHD simulations (Buck, 2020), these simulations show that such a population of stars is ubiquitous among these Milky Way-like galaxies.

In Section 3.6.2, we speculated that P184+43, with its planar orbit and unusual chemistry ([Mg/Ca]> +0.8), may have formed in a dwarf galaxy that was accreted at early times. The age estimate is based on its low metallicity [Fe/H]=−3.2 and chemical pattern, which suggests a lack of contributions from lower mass stars and SN Ia. It is possible that this star and its host dwarf galaxy were amongst the original building blocks that formed the proto Milky Way, as suggested from an analysis of NIHAO hydrodynamic simulations by Sestito et al. (2020a).

The chemistry of the Gaia-Sequoia candidates P016+28, P021+32, and P031+27 are largely consistent with previous studies, which show typical α-enhancement at low metallicities (Matsuno et al., 2019; Monty et al., 2020; Limberg et al., 2020; Yuan et al., 2020), though the large Mg-enhancement of P021+32 is higher than any previously observed Sequoia member. Similarly, Cordoni et al. (2020) find typical α-abundances, comparable to those seen in the Galactic halo, for their Sequoia candidates, which span −3.6 <[Fe/H]< −2.4. However, our other two candidates, P133+28 and P182+09, have sub-solar [Mg/Fe], and [Ca/Fe] lower than typical halo stars. Monty et al. (2020) also found one star in their sample of Gaia-Sequoia candidates (G184-007) with sub-solar [Mg/Fe] and low [Ca/Fe] ratios, though at a much higher metallicity ([Fe/H]=−1.67 for G184-007 vs. [Fe/H] = −2.79 and −2.85 for P133+28 and P182+09, respectively). If the location of the "knee" in the [Mg/Fe] vs. [Fe/H] plane, as identified by Monty et al. (2020) at [Fe/H]=−1.6 or −2.3, is an accurate reflection of the chemical evolution of Gaia-Sequoia, then the [Mg/Fe] ratios seen in P133+28 and P182+09 are difficult to reconcile as Gaia-Sequoia members. Alternatively, Gaia-Sequoia may have had episodes of star formation that stochastically enriched its interstellar medium, providing a range in some elements, like Mg. Similar abundance patterns have been seen in EMP stars in dwarf galaxies, such as Carina and Sextans (Norris et al., 2017; de Boer et al., 2014; Theler et al., 2020; Lucchesi et al., 2020). Furthermore, two of our Gaia-Sequoia candidates (P021+32 and P133+28) may be enriched in potassium (see Figs. 3.15 and B.2). An anti-correlation of stars that are Mg-poor but K-rich was discovered in the outer globular cluster NGC 2419 (Cohen et al., 2011; Cohen & Kirby, 2012). Other elements (Sc, and to a lesser extent Si and Ca) also showed variations, such that a more detailed analysis of these two Gaia-Sequoia candidates, especially with broader spectra that can reach more elements, would be interesting.

Finally, one of our new CEMP candidates, P016+28, appears to be *dynamically* associated with the Gaia-Sequoia event. We note that Yuan et al. 2020 identify one CEMP-no candidate with Gaia-Sequoia (CS29514-007, from Roederer et al. 2014). CS29514-007 is more metal-rich than P016+28, ([Fe/H]=−2.8 vs. [Fe/H] = −3.1) and has [Ba/Fe]~ solar, compared to the clearly neutron-capture rich P016+28 (we also measure La and Nd in this one star in our sample; see Appendix B.1.2). It would be interesting if these two stars probe the most metal-poor regime of the host of this event (also see the discussion on EMP stars in Gaia-Sequoia by Monty et al., 2020).

3.7 Conclusions and Future Work

We present detailed spectral analyses for 30 new metal-poor stars found within the Pristine survey and followed-up with Gemini GRACES high-resolution spectroscopy. All of these stars were previously observed with INT medium-resolution spectra.

- We confirm that 15 of our targets are EMP with [Fe/H]< −3.0 (50%), three of which only have iron upper-limits. If we consider their 1σ(Fe) errors, then we confirm 21 are EMP stars (70%). The most metal-poor star in the sample is P237+12, with [Fe/H]< −3.88.

- The INT medium-resolution spectra showed that three of our targets may be carbon-enhanced. We find that one is a CEMP-no candidate based on low [Ba/Fe] upper-limits, while the other two appear to be CEMP-s candidates.

- Two stars are found to be deficient in Ca, yet Mg enriched, yielding [Mg/Ca]=+0.8. This is a rare abundance signature, interpreted as the yields from a small number of SN II that underproduce Ca in explosive alpha-element production compared to Mg from hydrostatic nucleosynthesis.

- Five stars (P184+43, P191+12, P237+12, P244+10, P339.1+25.5) orbit in the Galactic plane, including our most metal-poor star (P237+12). We suggest they were brought in with one or more dwarf galaxies that were building blocks that formed the Galactic plane. As additional support one of these stars (P244+10) has a retrograde planar orbit. This star overlaps in eccentricity and action with the Gaia-Sequoia accreted dwarf galaxy; however, its low energy is in better agreement with the Thamnos event.

- Five stars are new candidates for the accreted stellar population from Gaia-Sequoia, based on their positions in the eccentricity-action-energy phase space (P016+28, P021+32, P031+27, P133+28, and P182+09). We find that P016+28, P021+32, P031+27 are enhanced in Mg and Ca, consistent with previous chemical studies of the Sequoia population; however, P133+28 and P182+09 both show low [Mg/Fe]. This could imply these stars are non-members; alternatively, we also find that P133+28 and P021+32 are K-enriched, and the Mg-K anti-correlation seen in P133+28 could indicate inhomogeneous mixing of the interstellar medium of its progenitor. We have found that P016+28 is also a CEMP-s candidate, showing enhancements of Ba, La, and Nd.

This work shows that the Pristine survey has been highly successful in finding new and interesting metal-poor stars, especially when combined with Gaia DR2 parallaxes and proper motions for testing stellar evolution and galaxy formation models. We look forward to the upcoming large spectroscopic surveys that will be able to tackle statistically large samples of these metal-poor stars for a detailed chemo-dynamical evaluation of the metal-poor components of our Galaxy.

Considering the Gemini/GRACES Large and Long Program is still ongoing, its' success thus far in uncovering new and interesting very metal-poor stars is remarkably promising. These results warrant future enthusiasm and the novel stars identified so far are prime candidates to be observed with the upcoming Gemini GHOST spectrograph (Chene et al., 2014; Sheinis et al., 2017). The excellent throughput expected at blue-UV wavelengths will allow for a more thorough assessment of stellar parameters and more detailed chemical abundance profiles.

Chapter 4

Spectroscopic Data Products

4.1   *Surpassing Imperfect Continuum Normalization in Medium Resolution Stellar Spectra*

Originally published in the Proceedings of the SPIE, Volume 10707, id. 107072W, 11 pp. (2018). Authors: Collin Kielty, Spencer Bialek, Sébastien Fabbro, Kim Venn, Teaghan O'Briain, Farbod Jahandar, Stephanie Monty

Personal Contributions

The machine learning architecture adopted for this work is the convolutional neural network (CNN) StarNet (Fabbro et al., 2018). In this investigative project, I expanded StarNet to be applied to medium-resolution stellar spectra and to spectra with poor continuum normalization. My contributions to the project, with guidance from the other co-authors, were in generating these data augmentations, creating figures 4.2 - 4.7, and writing most of the text except excerpts from section 4.1.1 and the "Signal-to-Noise Ratio" subsection of section 4.1.5.

4.1.1 Introduction

Our increasingly technological world is generating data in massive quantities unsuitable for traditional analysis methods. Solving the problem of efficiently analyzing large datasets has resulted in research into machine learning methodologies which, by their very nature, *require* an abundance of data. One of the most widely adopted and studied machine learning systems is the artificial neural network, due in large part to its strength as a universal approximator, capable of representing a wide variety of functions (Csáji, 2001).

The use of artificial neural networks in astrophysical applications is not a novel concept. Von Hippel et al. (1994) and Singh et al. (1998) implemented early machine learning models to drive research in stellar classification while Bailer-Jones et al. (1997) and Bailer-Jones (2000) adopted an artificial neural network to predict stellar parameters from synthetic stellar spectra.

More efficient, complex, and usable machine learning architectures have been developed, as computing power continues to increase, and large data sets become ubiquitous, leading to rapid advancements in astrophysical data processing. I refer the curious reader to Lee et al. (2008), Bensby et al. (2014), Holtzman et al. (2015), Ness et al. (2015), Casey et al. (2016), Pérez et al. (2016), Rix et al. (2016), Ting et al. (2016, 2017a,b, 2018, 2019), and El-Badry et al. (2018b) for an overview of the myriad of neural network applications, and their limitations, currently being implemented in stellar astrophysics.

Spectroscopic surveys like **SEGUE** (Yanny et al., 2009), **RAVE** (Kordopatis et al., 2013), **Gaia-RVS** (Recio-Blanco, 2016), **LAMOST** (Luo, 2015), **APOGEE** (Majewski & SDSS-III/APOGEE Collaboration, 2014), **GALAH** (De Silva et al., 2015) and **Ga-iaESO** (Smiljanic et al., 2014; Hawkins et al., 2016) provide high-quality homogeneous databases of stellar spectra for 10^5-10^6 stars. With upcoming spectroscopic surveys like **DESI** (DESI Collaboration, 2016), **4MOST** (de Jong et al., 2016), **WEAVE** (Dalton et al., 2016), **MOONS**(Cirasuolo et al., 2014), **SDSSV**(Kollmeier et al., 2017) and **MSE** (Mc-Connachie et al., 2016), fast and reliable analysis pipelines are needed to maximize the scientific impact of these studies. The ability of neural networks to process and analyze large sets of data quickly and accurately makes them an excellent application for these large-scale spectroscopic surveys that presently drive the field of Galactic archeology.

In this chapter, I present a new application of a convolutional neural network applied to the analysis of stellar spectra, *StarNet* (Fabbro et al., 2018). Several cases of training on synthetic spectra with StarNet are examined, predicting stellar parameters, and assessing their theoretical precision. Following the supervised learning approach, a representative subset of synthetic stellar spectra with known stellar parameters is selected and divided into a training set, or reference set, and a test set. The training set serves to constrain the neural network's mapping function from spectra to stellar parameters (effective temperature (T_{eff}), surface gravity ($\log g$), and metallicity ([Fe/H])), while the test set is used to asses the accuracy of the predictions of stellar labels. With a trained model, stellar parameters can then be predicted for the remainder of the sample. Our machine learning methods are summarized in Sec. 4.1.2 and our original StarNet model trained on high-resolution ($R \sim 20000$) synthetic spectra is presented in Sec. 4.1.3. In Sec. 4.1.4 StarNet is applied to medium-resolution ($R \sim 6000$) synthetic spectra and the precision in the stellar parameters is compared to what is achievable from high-resolution spectra. Sec. 4.1.5 is a preliminary investigation of the impact of continuum normalization and signal-to-noise on the stellar parameters.

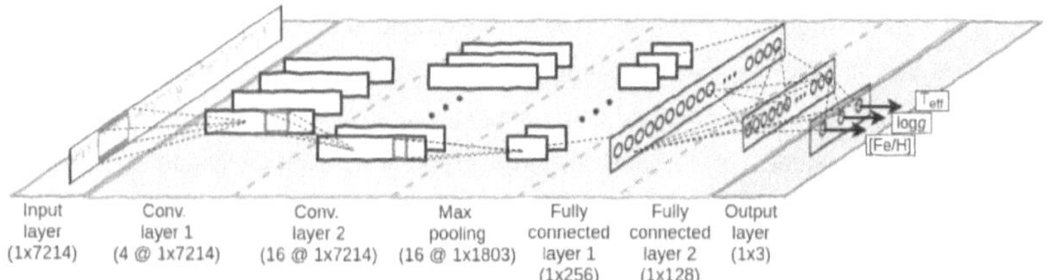

Figure 4.1 The StarNet CNN model reproduced with permission from Fabbro et al. (2018). The first layer is solely the input data followed by two convolutional layers with 4 and 16 filters (in successive order), then a max pooling layer with a window length of 4 units followed by 3 fully connected layers with 256, 128, and 3 nodes (again, in successive order). The last layer is the output layer.

4.1.2   The StarNet Convolutional Neural Network

In any supervised machine learning task, the goal is to learn a function which maps input data to a desired output, where the input and output can be scalars, vectors, or matrices. The flexibility of the input an output data permits a wide range of problems including regression, binary classification, and categorical classification. The mapping function can take on different forms, depending on the particular methodology adopted, but what sets machine learning apart from other methods is that the function is tuned with a dataset, and generalized to be applicable to additional datasets. The machine learning system adopted for this work is the convolutional neural network (CNN). For a full discussion of the StarNet CNN and our methodology please see Fabbro et al. (2018); the architecture is reproduced with permission from Fabbro et al. (2018) in Fig. 4.1 and the key points are briefly summarized below.

The StarNet CNN is composed of layers of neurons with trainable weights and biases. Each neuron takes inputs from the previous layer and performs a dot product followed by a non-linear transformation, resulting in an output for the neurons in the next layer to receive. There are two main types of layers present in the StarNet CNN: fully connected layers and convolutional layers. Fully connected layers are the classical neural network layers, in which each neuron uses the output from every neuron in the previous layer, that apply weights to each input value to compute an output, The convolutional layers see a localized view of the input and aggregate the data across a layer and assign importance to local features. A collection of *feature maps* are produced during the training phase which link these active filters to specific features within the spectra. Two successive convolutional layers are used in StarNet, allowing for the second of the two to convolve across the previous layer's feature map, and enabling the model to learn higher order features. The learned data are then down-sampled using maxpooling layers. These layers decrease the number of free parameters and extract the strongest features from the feature maps, while ignoring sections of the spectra that do no contain useful information. This reduces the amount of time required for each forward propagation as well as the time required for the model to reach convergence.

The combination of convolutional and fully connected layers allows the model to find relationships between individual flux values in a spectrum and the output stellar parameters as well as correlations between sections of the spectrum and the stellar parameters. This technique strengthens StarNet's ability to generalize its predictions on spectra of a wide range of signal-to-noise (S/N) and across a larger stellar parameter space.

4.1.3   StarNet on High-resolution Synthetic Spectra

The *a priori* stellar parameters, coupled with the ability to alter S/N ratios, resolution, and wavelength regimes make synthetic spectra a perfect tool to assess the accuracy and precision of StarNet. The original StarNet architecture in Fabbro et al. (2018) was trained on the high-resolution ($R \sim 20,000$) synthetic spectra used by the APOGEE consortium for the ASPCAP pipeline (Pérez et al., 2016). The "ASSET" (Koesterke et al., 2008) spectra were generated using both MARCS and ATLAS9 model atmospheres, as described Mészáros et al. (2012). The 6D ASSET grid is publicly accessible in a Principal Components Analysis compressed format and has been adopted as the training set for StarNet (see table 4.1 for the parameter coverage of the grid). Using a third order interpolation routine between spectra within the existing grid, spectra may be generated at any desired location in stellar parameter space.

To expand the predictive capabilities of StarNet at lower S/N ratios, Gaussian noise is added to all synthetic data, in both training and test sets. $S/N \approx 20$ up to near noiseless spectra was simulated, characteristic of the S/N ratios observed within the APOGEE survey (see Section 5.3 for further comments on the impact of the signal-to-noise ratio).

The synthetic dataset used in the high-resolution study, and paralleled in the later sections of this work, is comprised of 400,000 ASSET spectra produced through random sampling of stellar parameters within the ASSET grid. 232,000 spectra were randomly selected from our synthetic dataset to be used as the reference set. From the reference set, 192,000 (83%) were used to train StarNet, while the remaining 40,000 spectra were utilized

Table 4.1 Stellar parameter distribution of the ASSET synthetic spectra grid

Class		T_{eff}	$\log g$	[M/H]	[C/M]	[N/M]	[α/M]
	Min.	3500	0	-2.5	-1	-1	-1
GK	Max.	6000	5	0.5	1	1	1
	Step	250	0.5	0.5	0.25	0.5	0.25
	Min.	5500	1	-2.5	-1	-1	-1
F	Max.	8000	5	0.5	1	1	1
	Step	250	0.5	0.5	0.25	0.55	0.25

in the cross-validation procedure for the model at each iteration during training.

A test set of 40,000 spectra was selected from the remaining spectra not used in the reference set. Fig. 4.2 shows the residuals between the StarNet predictions and the *a priori* stellar labels assigned to the ASSET spectra. The variance of the residual distributions for all three predicted stellar parameters is well within the expected systematics expected from a classical "by-hand" analysis of stellar spectra. The model does show a clear dependence on S/N with higher degrees of scatter at lower S/N. StarNet also has the inclination to over-predict metallicity for low S/N and for spectra with [Fe/H]<-2.0. As absorption features disappear in IR spectra with [Fe/H]<-2.0, noise becomes the dominant spectral structure, biasing StarNet toward similar values of [Fe/H] for all very metal-poor stars.

4.1.4 StarNet on Medium-resolution Synthetic Spectra

Medium-resolution ($R \lesssim 10,000$) spectroscopic surveys like RAVE (Kordopatis et al., 2013) (R~7,000), SEGUE (Yanny et al., 2009) (R~2,000), LAMOST (Luo, 2015) (R~1,800), and Gaia Radial Velocity Spectrometer (RVS) (Recio-Blanco, 2016) (R~11,500) have proven to be exceptionally valuable for studies of structure within the Galaxy. Since spectrographs with high spectral resolution require significantly longer exposure times than a medium-resolution instrument, to reach the same desired S/N, medium-resolution surveys are typically capable of observing more targets, over a larger range of magnitudes, in a shorter period of time. From medium-resolution spectra of $> 10^6$ stars, the metallicity gradient of the Galaxy has been explored (Schönrich & Binney, 2009; Grand et al., 2015; Kawata et al., 2017), age-metallicity relations have been identified and linked to Galactic kinematic history (Martig et al., 2014; Aumer et al., 2016; Grand et al., 2016),and the metallicity-distribution function of the Galaxy has been mapped (Casagrande et al., 2011; Hayden et al., 2015). While detailed chemical abundances are historically inaccessible from medium-resolution spectra as a consequence of spectral feature blending at medium-resolution, the derivation of fundamental stellar parameters is still crucial for Galactic astrophysics.

Here StarNet is expanded to investigate its capability at medium-resolution. $R = 6000$ was chosen for the medium-resolution model to emulate the stellar spectra collected by the Gemini Near-Infrared Integral Field Spectrometer (NIFS) (McGregor et al., 2003), and the spectra expected from the proposed Gemini Infrared Multi-Object Spectrograph (GIRMOS) (PI Sivanandam, U. Toronto). Using the same methods outlined in Sec. 4.1.3, a training set of 192,000 synthetic spectra were randomly selected from the ASSET grid. Prior to adding Gaussian noise, the ASSET spectra were convolved with a Gaussian kernel to

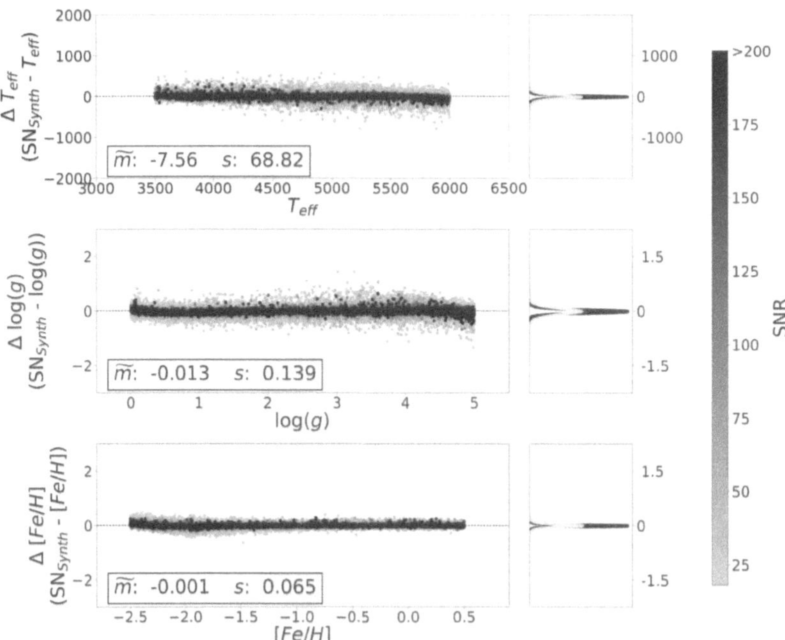

Figure 4.2 StarNet prediction residuals with *a priori* stellar labels for a test set of 40,000 high-resolution ($R \sim 20,000$) ASSET synthetic spectra (Koesterke et al., 2008). StarNet was trained with 192,000 synthetic spectra randomly sampled from the ASSET synthetic grid. Projected residual distributions are shown on the right (black for synthetic spectra with S/N > 80, gray for S/N < 60). The median value ($\widetilde{m}$) and standard deviation (s) are calculated in each panel.

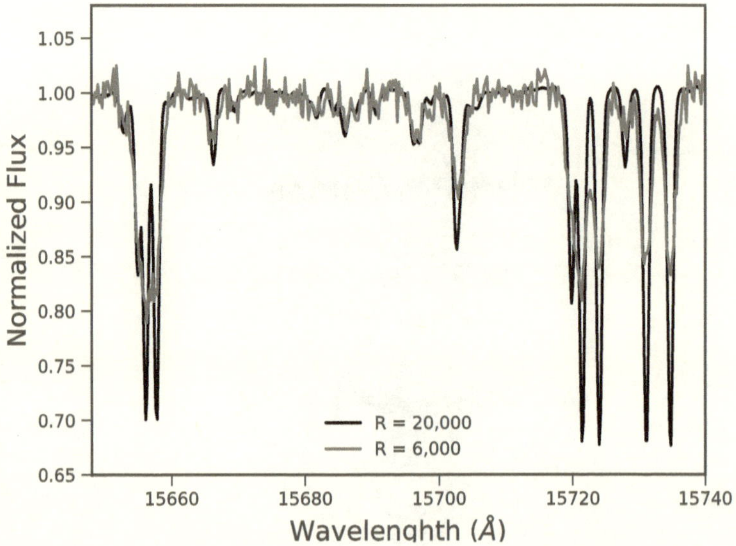

Figure 4.3 A sample region of a high-resolution ($R \sim 20,000$) ASSET spectrum (black) and the corresponding noise-added, medium-resolution ($R = 6,000$) spectrum (red). The combination of line blending (particularly noticeable around the CO feature at 15725 Å) and noise introduces uncertainty in the continuum placement, an effect explored in detail in Sec. 4.1.5.

degrade the resolution from $R \sim 20,000$ to $R = 6000$ (see Fig. 4.3 for a sample spectrum). Cross-validation and test sets of 40,000 spectra were prepared using the same technique.

Similar to Fig. 4.2, Fig. 4.4 shows the residuals between the StarNet predictions and the known labels from ASSET spectra, but for the model trained and tested on $R = 6000$ spectra. Comparing the two figures, StarNet systematically over-estimates T_{eff} by ~ 60K and $\log g$ by ~ 0.13 dex when the predictions are derived from a medium-resolution model. Furthermore, a higher degree of scatter is seen in the T_{eff} residuals when T_{eff} is derived from medium-resolution spectra (~ 90K vs. ~ 70K from high-resolution spectra).

Though these small offsets have implications about the limitations of medium-resolution spectra (see Sec. 4.1.5), the predictions from high-resolution and medium-resolution

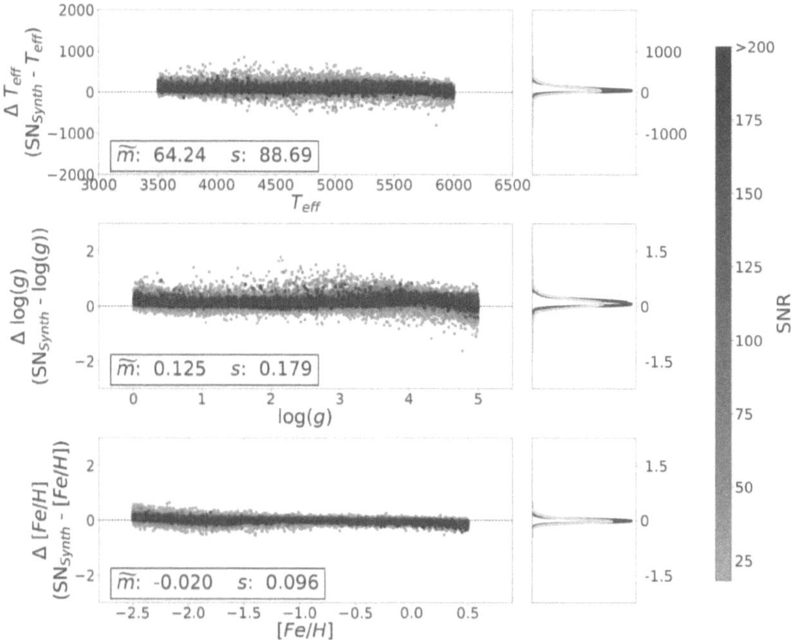

Figure 4.4 Same as Fig. 4.2 but for a StarNet model trained on 192,000 medium-resolution ($R = 6,000$) ASSET synthetic spectra. The test set shown contains 40,000 medium-resolution spectra prepared in an identical manner to the training set.

models are in excellent agreement with each other. This is encouraging for the use of StarNet as fast and accurate tool in the analysis of medium-resolution IR spectra.

4.1.5 Impacts of Continuum Normalization and Signal-to-Noise Ratio

Continuum Normalization

In Sec. 4.1.4, it was shown that StarNet can be trained on medium-resolution IR spectra and can make predictions for stellar parameters with similar precision to labels derived from a high-resolution analysis. Despite the agreeable precision, small systematic offsets are seen in the residuals of the predictions on T_{eff} and $\log g$. This may indicate a bias in the medium-resolution training set that is not present in the high-resolution data. In this

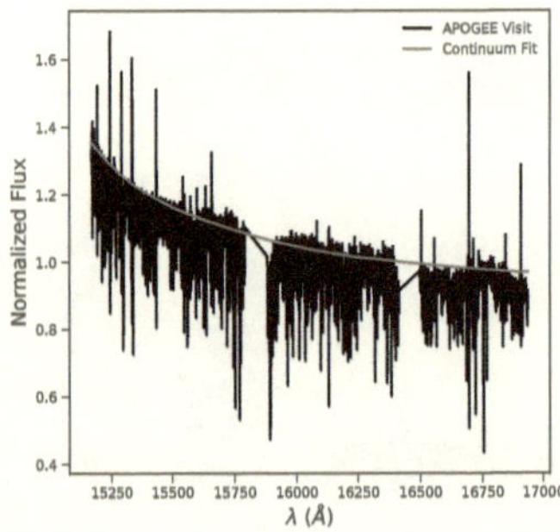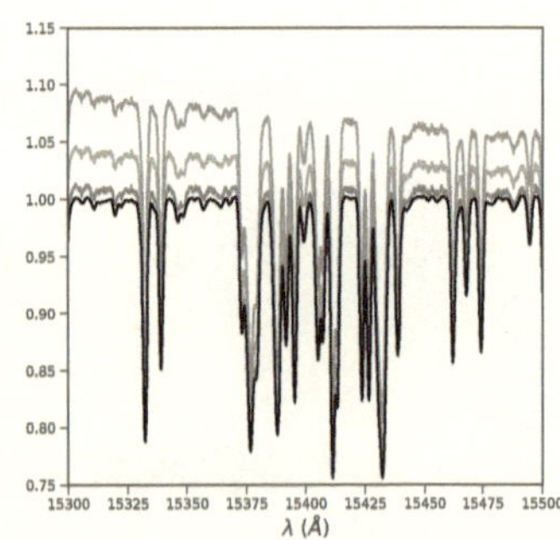

Figure 4.5 Left panel: A characteristic normalized APOGEE visit spectrum (black) with a seventh-order polynomial fit to estimate the placement of the continuum (blue). Right panel: A sample region of a medium-resolution ASSET spectrum (black). The continuum from the left panel was rescaled and added to the ASSET spectrum to emulate an "APOGEE-like" signature with varying degrees of continuum error. The blue/orange/green spectra corresponds to 1%, 4%, 8% offsets in the continuum.

section, I investigate whether these biases are the result of continuum placement and attempt to quantify the effect on the StarNet results.

In high-resolution spectra, continuum is determined by fitting a low order polynomial to spectral regions that are free of strong absorption lines (Ting et al., 2019; Ness et al., 2015; Pérez et al., 2016). This is complicated in medium-resolution spectra though as line blending can obscure the location of the true continuum. When the polynomial structure is removed from the spectrum, spectral features can appear weaker than their true strength[1]. Since T_{eff} is crucial in determining line strength, particularly in IR spectra, weaker lines can either indicate a hotter object or a more metal-poor star. I examine the predictive power of StarNet by testing on spectra with imperfect continuum removal.

A library of ~993,000 polynomials was created by fitting a seventh order polynomial to all individual visit spectra from APOGEE DR14 with the goal of generating continuua characteristic of real observed spectra (see Fig. 4.5 for a sample APOGEE spectrum and the corresponding continuum fit).

The StarNet model described in Sec. 4.1.4 was trained on synthetic spectra, therefore the continuum was known *a priori*, then convolved with a Gaussian to reduce the resolution to R ~6,000. Recall in Sec. 4.1.3 and Sec. 4.1.4 the effect of S/N on the StarNet results is

---

[1] A similar effect is seen in metal-poor stars, even at high-resolution

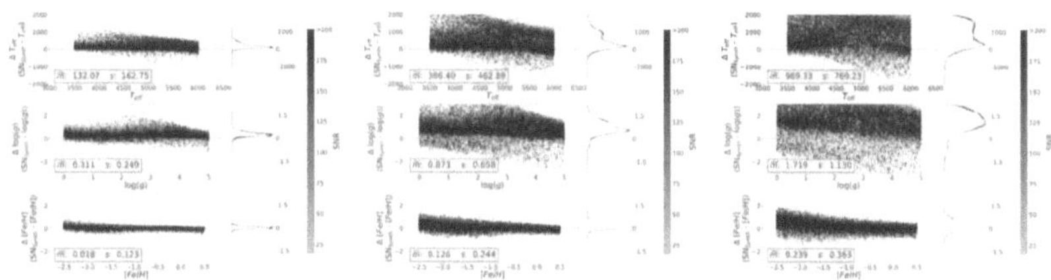

Figure 4.6 Same as Fig. 4.4 but for a StarNet model tested on synthetic spectra with continuuum offsets. The left/middle/right panels are the predictions for a model tested on spectra with a 1%/4%/8% continuum offsets. The precision in the StarNet results rapidly degrades with even small continuum offsets.

examined. In addition, I explore the effect of assigning a real "APOGEE-like" continuum randomly sampled from the library of generated polynomials. Those spectra were then rescaled to reflect a certain degree of error in the continuum placement. The error in the continuum placement was then defined as the standard deviation of the normalized flux away from the true mean of 1.0 (known from the original synthetic spectra). The right panel in Fig. 4.5 shows a sample of an ASSeT spectrum, at $R \sim 6,000$, with and average 1%, 4%, and 8% offsets in the continuum.

StarNet was then trained on these medium-resolution spectra with continuum offsets. Medium-resolution spectra with 1% continuum offsets are shown in Fig. 4.6 (left panel): both a larger scatter and systematic offset are seen in the stellar parameters. Both are a factor of $\sim$2 larger compared to the stellar parameter residuals seen in Fig. 4.4.

StarNet systematically over-estimates T_{eff} and $\log g$ as a result of the weakened spectral features. As seen in the middle and right panels of Fig. 4.6 the impact is even larger when 4%, and 8% continuum offsets are introduced. These tests serve as a reminder of the importance of continuum normalization in determining precision stellar parameters, and the challenge of ascertaining the continuum from real data at medium-resolution, e.g. for observations taken with NIFS and GIRMOS.

StarNet Adaptability to Continuum Normalization Errors

The previous section highlights the limitations of StarNet due to continuum offsets. However, if StarNet is trained on synthetic spectra with continuum offsets, do the predictions improve? To test this, a training set of 192,000 $R = 6,000$ synthetic spectra and a cross-validation set of 40,000 synthetic spectra were generated from the ASSET grid. A randomly

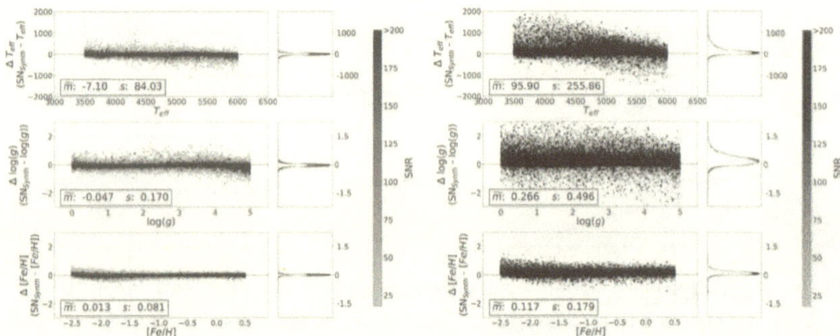

Figure 4.7 StarNet predictions for stellar labels from a model trained on 192,000 medium-resolution ($R = 6,000$) ASSeT synthetic spectra with added"APOGEE-like" continuum errors. The offset in the continuum for each spectrum in the training set was randomly sampled from a normal Gaussian distribution with a standard deviation corresponding to a 5% offset. The left panel shows the theoretical limit of the model when tested on medium-resolution spectra with no added continuum offset errors (as in Fig. 4.4). The right panel shows the predictions when trained and tested on spectra with continuum offsets.

selected "APOGEE-like" continuum was added to each spectrum in the training and cross-validation sets. These were then rescaled such that the average offset was less than or equal to 5%. To be more specific, the rescaling factor was sampled from a normal Gaussian distribution with a standard deviation corresponding to a 5% offset.

Once trained, this model was tested on two different test sets of 40,000 medium-resolution synthetic spectra: one containing no offsets in the continuum (as in Sec. 4.1.4) and another prepared in the same fashion as the training set. The former test set demonstrates the theoretical limit of the model, while the latter tests whether the errors in continuum offset can be reduced or removed by StarNet. The results of these tests are shown in Fig. 4.7.

Comparing the left panel of Fig. 4.7 to the middle panel of Fig. 4.7, a factor of 6-3 improvement is seen in the residuals for T_{eff}, $\log g$, and metallicity for the model trained on spectra with continuum offsets. In fact, when comparing the left panel of Fig. 4.7 to Fig. 4.2, these results imply medium-resolution spectra, with continuum offset errors, can be used to train StarNet to similar precision as high-resolution spectra (when there are no continuum offset errors in the test sets). Fig. 4.7 right panel shows that the challenge really is the continuum normalization of the medium-resolution data in the test set.

The Signal-to-Noise Ratio

A commonly adopted method of estimating the continuum is via asymmetric sigma clipping; a spectrum is split into several segments, each of which is iterated over to reject points that are discrepant by a specified number of standard deviations from the median, with the points below being more aggressively rejected. Once the continuum level is determined in each segment, an n-th order spline is fit to the continuum points. This method has a couple distinct advantages: 1) given an appropriate segment length (a common choice is 10Å), typically the majority of data points will be closer to the continuum and thus absorption features will be rejected in the fit, and 2) instrumental signatures affecting the overall shape of the continuum in non-linear ways are relatively simple to capture, especially compared to other methods which attempt to fit an n-th order polynomial over the entire spectrum.

Of course, no method is without its own set of limitations. Due to more aggressively rejecting points below the continuum, an unintended consequence is that the envelope is pushed to the noise ceiling of the spectrum, above the true continuum (see Figure 4.8). When the spectrum is subsequently divided by the continuum, all absorption features will thus appear weaker. Using traditional methods of stellar parameter estimation involving fitting a grid of *noiseless* synthetic spectra to a *noisy* observed spectrum, all of which were continuum normalized in the same way, this unaccounted for bias in noise can lead to systematic errors. In StarNet, training on theoretical spectra with Gaussian noise helps to avoid these systematic errors.

Furthermore, as long as the training dataset of a NN is properly and carefully chosen, its predictions can be generalized to a dataset that it hasn't seen yet. Often times this includes augmenting the training dataset with features that approximate the features of the dataset you wish to make predictions on. In the case of StarNet being trained on synthetic spectra this means modifications due to: a) rotational velocity, b) radial velocity, c) macro/microturbulence, d) degradation to instrumental resolution, and e) the addition of noise. StarNet can learn to predict values based on the presence of these features (the first three) or learn to ignore them in its predictions through appropriate training (the last two).

4.1.6 Conclusion

I have demonstrated that a CNN model is capable of determining the stellar parameters T_{eff}, $\log g$, and [Fe/H] directly from both high-resolution and medium resolution stellar spectra. By applying StarNet directly to synthetic spectra, modified to simulate the data expected from IR spectrographs, StarNet is capable of estimating stellar parameters from a training

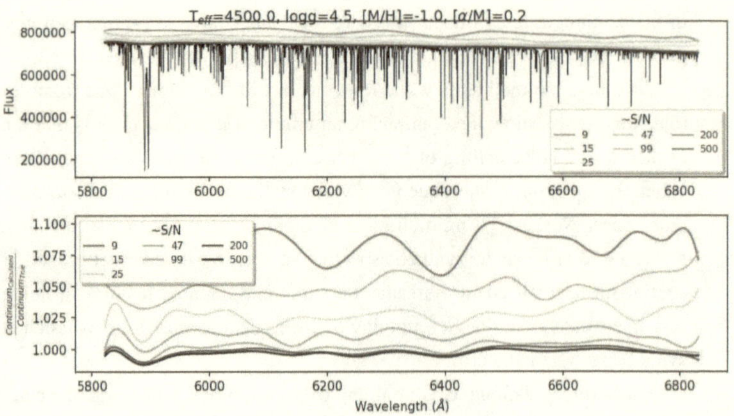

Figure 4.8 The results of asymmetric sigma clipping continuum fitting of an AMBRE spectrum which was modified by the addition of varying amounts of Gaussian noise.

set of synthetic spectra alone. In the high-resolution analysis, the uncertainties in the stellar parameters derived by StarNet are comparable to, if not better than, what is expected from a conventional analysis, over a large range of S/N ratios.

Using medium-resolution spectra, StarNet is also able to recover stellar parameters to equal precision as the high-resolution model, when effects such as continuum normalization errors are ignored. However, the precision in the stellar parameters from medium-resolution spectra rapidly degrades when a simple continuum error model is incorporated into the test set. While StarNet can improve the results if there are continuum normalization errors in the training set, the challenge remains to set up spectroscopic surveys which minimize continuum normalization errors in the data.

We plan to use StarNet to explore other aspects of automated spectral analyses, for example wavelength range, line profile variations, and intrinsic errors in the NN model, to prepare for the era of spectroscopic surveys (Bialek et al., 2020).

4.2   *A Data Reduction Pipeline for Gemini-North's Near-Infrared Integral Field Spectrometer*

Originally published in The Astronomical Journal, Volume 158, Issue 4, article id. 153, 15 pp. (2019). Authors: Marie Lemoine-Busserolle, Nathaniel Comeau, Collin Kielty, Kerry Klemmer, and Megan E. Schwamb.

Personal Contributions

Gemini's Near-Infrared Integral Field Spectrometer (NIFS; McGregor et al. 2003) is a medium-resolution ($R \sim 5300$) near-infrared instrument located at Gemini-N. As an Integral Field Spectrometer (IFS), NIFS provides spatially resolved spectra and is an excellent instrument to study gas dynamics around active galactic nuclei (Riffel et al., 2008; Storchi-Bergmann et al., 2010), to perform spatially resolved spectroscopy on gravitationally lensed galaxies (Livermore et al., 2015), to directly image exoplanets and other substellar companions (Lafrenière et al., 2010; Wahhaj et al., 2011), and to collect spectra for many stars simultaneously in crowded fields-of-view (Davidge et al., 2010, Turri et al. in prep.). It is this last point, in conjunction with my expertise in handling near-IR stellar spectra, that piqued my interest to intern at Gemini-N to work on the data reduction pipeline for NIFS as part of the New Technologies for Canadian Observatories (NTCO-CREATE) training program.

Section 4.1 demonstrated the resourcefulness of novel methods to *analyze* many stellar spectra when the data are already in hand, however the *homogeneous acquisition and reduction* of stellar spectra needed for machine learning algorithms is not trivial, especially when the data are from an IFS. The need for a flexible, efficient, and reproducible data reduction pipeline to reduce NIFS data and produce science ready 3D data cubes warranted the creation of *Nifty4Gemini*, an open-source, python based data reduction suite for Gemini NIFS data. Upon my arrival at Gemini, a majority of the *Nifty4Gemini* pipeline had already been written; however, there was (and still is) much room for improvement. Telluric correction is an important step in the science reduction for users who require precision measurements of individual spectral lines for abundance analyses and/or radial velocity studies. The original telluric correction routine queries SIMBAD (Wenger et al., 2000) to determine the spectral type, temperature, and 2MASS (Skrutskie et al., 2006) magnitudes of the telluric standard star. These values are passed to the IRAF task *iraf.telluric* to shift and scale a $R \sim 5000$ spectrum of Vega to create a theoretical spectrum of the standard star. This model spectrum is then divided from the observed telluric standard to remove the standard star spectrum of its intrinsic absorption lines. The shortcoming of this procedure, however, is that this process assumes an early A-type star was observed as a spectroscopic standard. I reworked Nifty to accept a reference/model spectrum specified by the user and/or to accept an already corrected telluric star spectrum. After intrinsic absorption lines are removed, a normalized telluric spectrum is produced. The original normalization method fit a third-order cubic spline to the telluric star spectrum and divided the telluric spectrum by this fit. Since broad lines, like the H lines in the K-band, and the telluric lines themselves affect the continuum placement, an asymmetric sigma-clipping routine was included prior to the spline fit to improve the fidelity of the continuum estimation.

I additionally created a prototype of a "*Nifty4Gemini-lite*" that would exist within Gemini's internal data archives. The aim was to reduce NIFS data the night it was collected and produce "quick-look" data cubes for PI's to assess their observing strategies, instrument set-up, and predicted vs. measured SNR. The Gemini internal version of Nifty queries the Gemini data archives for new NIFS data each night, sorts the data based on the Gemini program ID, and cross-matches with the relevant calibration frames (which might be used for multiple programs). Based on the completeness of the calibration files found, the pipeline will reduce the observed data to the highest level it can, stopping short of merging multiple cubes due to the possibility of non-sidereal targets or unique observing specifications. A reduction summary log is created and emailed to the Principle Support astronomer assigned to the particular program ID. If the reduction meets the Gemini data quality assessment standards, the cubes are made available for PI's. At the time of my departure from Gemini, the data quality assessment team was working on updating their tools to handle NIFS data cubes so the Principle Support astronomer can be removed from the process and the cubes can be accessed online by PI's directly.

My final contributions to the *Nifty4Gemini* project was in writing the Lemoine-Busserolle et al. (2019) paper. I created all figures, tables, and associated captions excepting Figures 1, 10, 11, and 12, added substantially to the text, and addressed the questions posed in the first round of referee reports. The Lemoine-Busserolle et al. (2019) paper is reproduced in the following sections with my personal contributions in bold text.

ABSTRACT

I present a python package, called *Nifty4Gemini*, and its associated Pyraf/Python based pipeline for processing Gemini-North Near-Infrared Integral Field Spectrometer (NIFS) observations. Built on the Gemini IRAF package's capabilities, *Nifty4Gemini*'s associated NIFS pipeline is a data reduction package which reduces NIFS raw data and produces a flux and wavelength calibrate science cube with the full Signal/Noise (S/N), ready for science analysis. It utilizes tasks from the Gemini IRAF package, PyRAF and packages from the Gemini AstroConda environment. *Nifty4Gemini* is a configuration-based pipeline framework written in python which is easily extensible to integrate additional pipelines and user-defined scripts. *Nifty4Gemini* is open-source and available for download with its documentation available .

4.2.1 Introduction

Gemini's Near-Infrared Integral Field Spectrometer (NIFS; McGregor et al. 2003) is a facil-ity near-infrared spectrograph which supports seeing-limited, natural guide-star adaptive op-tics (AO), and laser adaptive optics (AO) near-infrared spatially resolved spectroscopy when used with the ALTtitude conjugate Adaptive optics for the InfraRed (ALTAIR; Richardson et al. 1998), the adaptive optics (AO) system on Gemini North (Christou et al., 2010). NIFS is currently available for use on the Gemini Observatory's 8.1-m Frederick C. Gillett Gemini North Telescope on Maunakea. The NIFS detector is a Rockwell HAWAII-2RG (H2RG) device with 2048x2048 18μm pixels. The outer four pixels on each side are not illuminated, and so act as reference pixels. This leaves an active area of 2040x2040 pixels.

The HAWAII-2RG detector is sensitive to light out to 2.6μm, and uses a HgCdTe detector layer. It uses four output amplifiers that simultaneously read out 512x2048 pixels in around 5 seconds. A recorded NIFS frame is actually the difference in signal between two read-outs of the detector: one at the start of the integration, and one at the end. For this reason, the minimum permitted exposure time is determined by the read-out time. At the heart of NIFS is a reflective integral-field unit (IFU) which divides its $\sim 3.0''$x 3.0$''$ field-of-view on the sky into 29 slitlets each 0.105$''$ wide and 3.0$''$ long. Spectra are obtained simultaneously for each 0.043$''$ pixel (69 on detector in total) along each slitlet. This results in rectangular spatial pixels (or 'spaxels') of dimension 0.103$''$x0.043$''$ across and along the slice respectively. Four reflection gratings are used with a fixed focal length camera to obtain spectra with two-pixel resolving powers of R $\sim$ 5300 in any one of the Z, J, H, or K bands.

In this chapter, I describe *Nifty4Gemini*, a python package, which includes a data processing pipeline for NIFS observations. The NIFS pipeline was developed to reduce NIFS raw data and produce a final flux and wavelength calibrated science data cube with the full signal to noise needed for science analysis. It is based on pre-existing Image Reduction and Analysis Facility (IRAF)[2] routines that are part of the Gemini IRAF Package[3], which is installed by AstroConda[4], and PyRAF[5]. The raw data from the Gemini facility instruments are stored as Multi-Extension Flexible (MEF) Image Transport System files (FITS). Therefore, all the tasks in the Gemini IRAF package, intended for processing data from the Gemini facility instruments, are capable of handling MEF files. IRAF, PyRAF and the Gemini IRAF Package have been widely used in the astronomy community and therefore have been extensively tested and are supported by Gemini Observatory. However those routines are not enough to create a pipeline, therefore they are used as a library of computer code for the NIFS pipeline, instead of re-writing the function themselves that already exist in IRAF and the Gemini IRAF Package. *Nifty4Gemini* and its NIFS pipeline are introduced as a software pipeline written in the python programming language, which is designed so it is easy to run and repeat the data reduction process. It is also easily extensible and straightforward to install.

In Section 2, I give an overview of *Nifty4Gemini*'s structure and the NIFS Pipeline. In Section 3, I describe the NIFS raw data and discuss the various steps of the data reduction

---

[2]IRAF is distributed by the National Optical Astronomy Observatory, which is operated by the Association of Universities for Research in Astronomy (AURA) under a cooperative agreement with the National Science Foundation.

[5]PyRAF is a product of the Space Telescope Science Institute, which is operated by AURA for NASA

that the NIFS pipeline performs as in the V1.0.1 release. In Section 4, I describe tutorials and example available in the online documentation of with *Nifty4Gemini*. Finally, conclusions and prospects for future development of *Nifty4Gemini* are presented in Section 5.

4.2.2 General *Nifty4Gemini* Overview

Our NIFS pipeline software is run via *Nifty4Gemini*, a python configuration-based framework which gathers input from the users, generate configuration files and control the workflow of the data reduction performed by the pipeline. The first release of *Nifty4Gemini* contains our NIFS pipeline and its multiple steps to carry out a data reduction (which by default are automated). In the *Nifty4Gemini* framework a "step" is defined as a major part of the data reduction and each step has its self-contained Python script. Therefore each step can also be run on its own. Each step is composed of various tasks. In our framework a task is an individual PyRAF or pure Python function that does much of the data processing of a Pipeline. Each task has input data, output data, input parameters and optional return values. *Nifty4Gemini* uses a configuration file reader and writer named *ConfigObj* (version 4.7.2) [6] to create and load a configuration file, called *config.cfg*. All data reduction parameters used by the NIFS pipeline are saved in the *config.cfg* file, which is stored and can be shared, making it easy to exactly reproduce a data reduction workflow. *Nifty4Gemini* is open source [7] and the git repository includes the source code with detailed installation instructions, a quick-start guide, and a link to the full documentation, which is hosted on ReadTheDocs [8].

4.2.3 NIFS Data Reduction Pipeline

I present here a typical NIFS data set and describe the major steps of our NIFS pipeline to reduce and calibrate such observations (Fig. 4.9).

Required Dataset

Most NIFS observations use a standard setup with science and associated sky frames as well as a basic set of frames for calibration measurements, which is referred to as baseline calibrations. The baseline calibration set usually includes : (1) Spectral Flat Field Frames (lamps-on and lamps-off frames), which are required for each grating setting in order

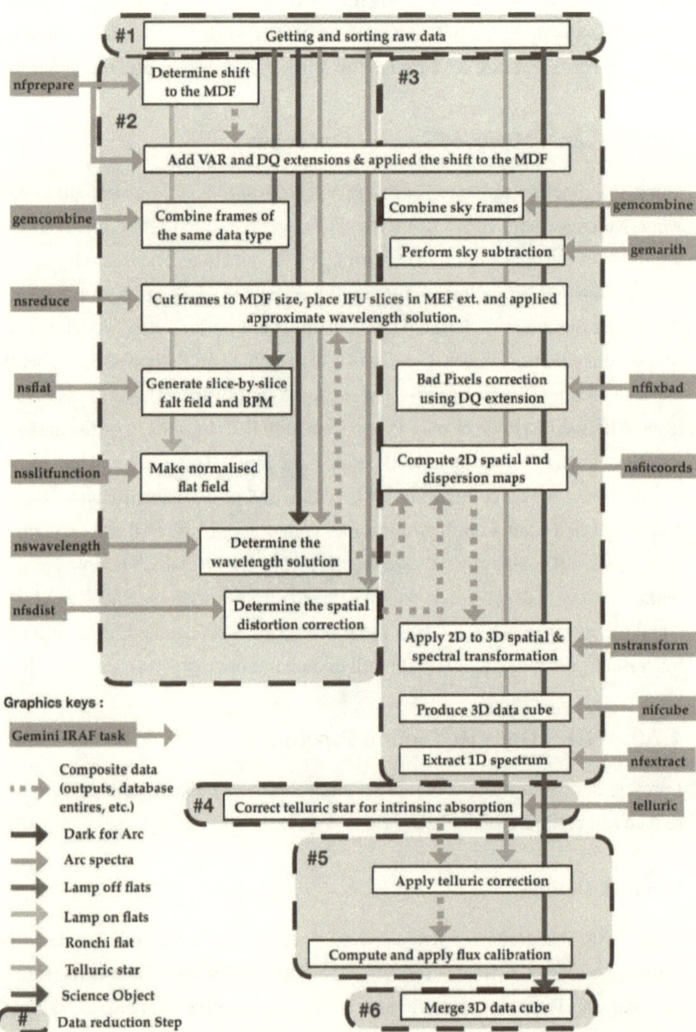

Figure 4.9 NIFS pipeline's workflow which includes the Gemini IRAF tasks used during the different steps. There is no Gemini IRAF task used for the last step of the data reduction

to correct pixel-to-pixel gain variations in the spectrograph detector and the wavelength-dependent throughput of the spectrograph (including the wavelength-dependent response of the optics, filter, grating, and detector). (2) Distortion Calibration Frames obtained using a coarse Ronchi grating, which are used to calibrate the spatial dimension of the IFU field. (3) Wavelength calibration frames, based on exposures of arc lamps, which are also obtained each time the grating turret in NIFS is moved. (4) Dark frames obtained in order to subtract the dark current pattern from the arc frame. (5) Four exposures on-source of a smooth spectrum star are generally obtained together with two sky frames of the same exposure time. Those frames can be used to correct telluric absorption lines and also used for flux calibration if that telluric star's broad-band near-infrared magnitudes are known. It is worth noting that each science data set and Telluric data set also comes with associated acquisition images, used to center the science target in the NIFS Field-Of-View (FOV). All NIFS raw MEF data files have two unnamed extensions. Most of the header information is written to the Primary Header Unit (PHU), extension [0]. The data read from the detector array is in extension [1].

Step 1 : Getting and Sorting the NIFS raw data

The raw NIFS observations can either be located in a local directory or the pipeline can download them from the Gemini Observatory Archive[9] (if the data are public) by providing a program ID (e.g., GN-2017B-Q-1). In the latter case, the raw data will be automatically downloaded, verified (by ensuring that a minimum of data was found for science target, calibration and telluric star), and decompressed. *Nifty4Gemini* uses some components of NDMApper[10] for this purpose. The first step of the data reduction sorts the raw NIFS data in appropriate directories according to their types, date obtained and grating configurations. During this step, using the information present in headers of each MEF files, the pipeline creates a new directory structure, text files with list of files names by types (e. g. calibrations, science target, telluric star, science target acquisition) and will write the paths to the newly created directories in the configuration file config.cfg. It is important to note that for NIFS there is no difference in the headers between lamps on flats (referred to as *flats*) and lamps off flats (referred to as *flatdark*). For all flats, the pipeline computes an average counts and if the median is greater than 2000, the image is identified as a flat and if it is less it is identified as a flatdark. **An example of raw calibration and science frames are shown in Fig.** **4.10.** **All data reduction figures use** K**-band observations of Titan (program**

GN-2014A-Q-85; obtained in Apr./May 2014) as an example.

The pipeline also at this stage associates science frames with telluric star frames that are closest in time (less than 1.5 hours) and stores this information in a text file to be used later in the data reduction process. This step needs to be run only once unless additional raw data are added. All tasks in this first step are written entirely in python.

Step 2 : Reducing the NIFS raw calibration data

The second step is to reduce the NIFS baseline calibration observations. This sequence is performed on the calibration data located in each calibration directories. Step 2 is composed of four tasks, each of them are a python wrapper around Gemini IRAF tasks. During the first task, *iraf.nfprepare* from the Gemini IRAF package is used to determine the image slicer offsets and to calculate variance and data quality frames. Headers values in the observation determine the observation mode and so identify a particular Mask Definition File (MDF). The MDF describes the illumination pattern on the detector of the image slicer and *nfprepare* uses it to compute any spatial shift of the IFU maps on the detector for each NIFS observation. The output is referred to as the "shift frame". In addition, using the information in the MDF, three extensions are extracted from the original raw data: SCI (the data), VAR (variance) and DQ (data quality). The generation of the data quality (DQ) plane is important in order to identify and fix hot and bad pixels on the NIFS detector in subsequent steps in the data reduction process. **Pixels with ADU (Analogue-to-Digital Unit) values that exceed the linear limit of the detector ($\sim$35,000 ADU), or pixels which are saturated ($\sim$48,000 ADU), are flagged within the DQ frame at this step in the reduction process. In addition, the use of *nfprepare* and *iraf.nffixbad* will minimize the effect of bright cosmic ray strikes. Cosmic ray effects are also diminished with *iraf.gemcombine* via median combination of multiple frames of the same data type, and *iraf.nifcube* will further suppress their effect by merging reduced data cubes from multiple science acquisitions (see Sec. 4.2.3).**

In the second task, all lamps-on and lamps-off flats are run through *iraf.nfprepare* with the shift frame used as a reference to apply the MDF shift. The variance and quality extensions are also generated. All lamps-on flats are combined into a master lamps-on flat using *iraf.gemcombine* and all lamps-off flats are combined in a similar fashion. *iraf.nsreduce* is then used on the two combined lamps-on and lamps-off flats frames (1) to cut the frames to the size specified in the MDF, (2) to place each IFU slice in separate MEF

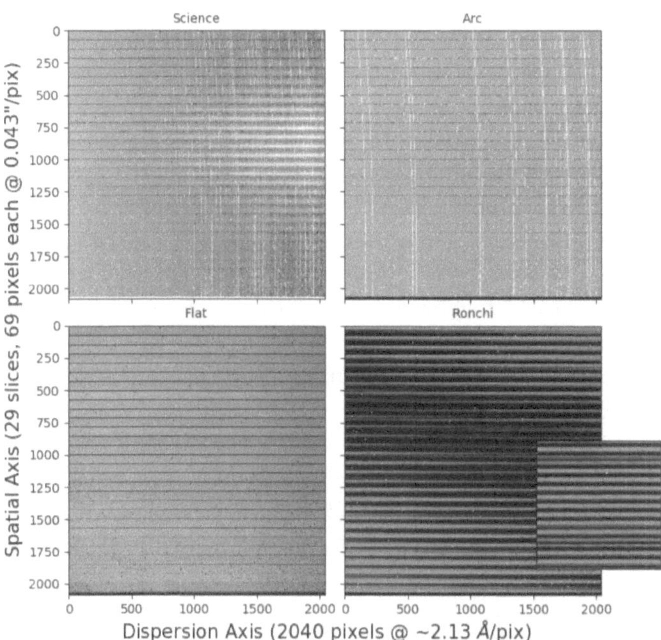

Figure 4.10 **Example of raw NIFS K-band science and calibration (lamp-on flat, lamp-on dark, and Ronchi) frames in the detector plane. Easily seen in the flat frame (*bottom left*), the dark horizontal lines outline the edges of each IFU slice. A zoom in region of the Ronchi frame is shown to display the Ronchi slit-mask (*bottom right*).**

extensions, (3) and to apply an approximate wavelength calibration[11]. In order to ensure the number of bad pixels caught is approximately constant for each observing band, the pipeline has different lower rejection threshold values for each grating in the creation of the BPM. **The process to select the optimal rejection threshold value for each band was as follows: 1) A set of DQ frames was produced for a given grating, utilizing a range of rejection threshold values. 2) The number of bad pixels identified in each frame, for a given threshold, is recorded and compared to other frames processed using the same threshold. 3) The threshold which rejects the most consistent number of pixels between frames is chosen for that grating (typical rejection count differences between frames are on the order of 10 pixels). 4) Steps 1-3 are repeated for all gratings. 5) The number of rejected pixels for each grating is compared and the thresholds are iteratively adjusted until the number of pixels rejected between gratings is comparable to the number of pixels rejected between frames of a particular grating (on the order of 10 pixel differences). Lower rejection thresholds, in terms of a fraction of the median value in the frame, are 0.07 for the Z and J gratings and 0.05 for the H and K gratings.**

The final flat field frame and BPM are created with *iraf.nsslitfunction* (Fig. 4.12) by correcting the normalization done by *iraf.nsflat* for slice-to-slice variations. The output from this task is used as the flat field image and BPM (Fig. 4.13) for further reduction.

In the third task, similarly as for the lamps-on and lamps-off flats, the arc lamp frame (called arc frame) and associated dark frame (called arc dark frame) are run through *iraf.nfprepare*, *iraf.gemcombine* (if there are more and one arc lamp and arc dark frame), after those frames are processed by *iraf.nsreduce* with the flat field being applied. The wavelength solution for each slice is obtained by running *iraf.nswavelength*, which uses the core IRAF tasks *identify/reidentify*.

***iraf.nswavelength* calls custom high resolution line lists designed for the Gemini Calibration Unit (GCAL) which are suitable for the spectral resolution of NIFS. These line lists have been vetted by eye to remove weak/blended lines that obfuscate automatic line identification. The line list which best covers the wavelength regime of the grating in use is loaded. Fig. 4.14 shows typical arc spectra for each of the four gratings with line list coverage overlaid. A line list summary and typical RMS errors on the wavelength solution for each grating are given in Table 4.2. Thresholding parameters**

---

[11]**This approximate wavelength solution is only needed within *iraf.nsreduce* and is not reflected in the science frames; proper wavelength calibration is done at a later stage.** The outputs are then run through *iraf.nsflat* to generate a normalized slice-by-slice flat field frame and a slice-by-slice BPM (Bad Pixel Mask) frame via thresholding (see Fig. 4.11)

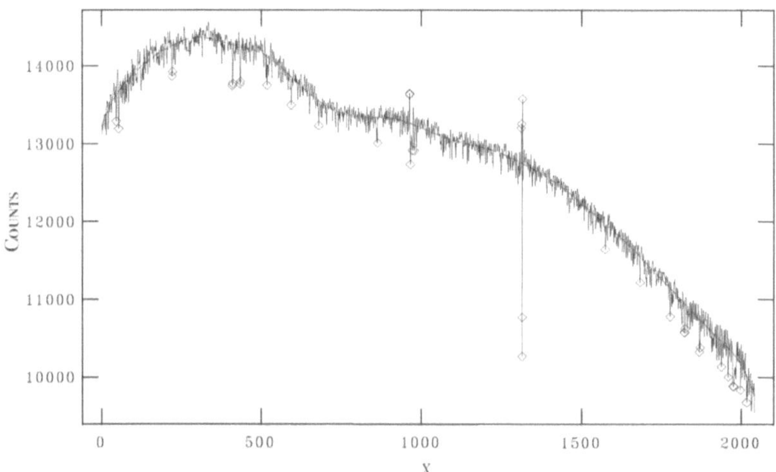

Figure 4.11 **Example of fitting a K-band lamps-on flat spectrum using *iraf.nsflat*. Pixels rejected via thresholding are identified by open diamonds and an order 20 cubic spline is fit to normalize the flat field.**

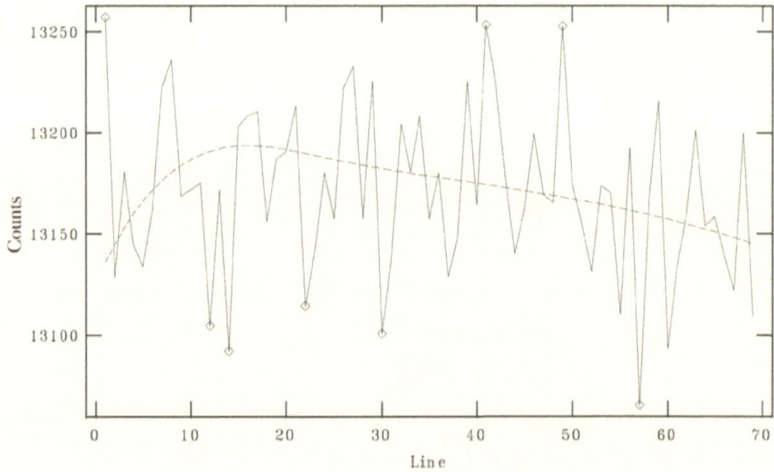

Figure 4.12 **Example of correcting the *iraf.nsflat* normalization for inter-slice (spatial) variations within a single *K*-band flat slice using *iraf.nsslitfunction*. Pixels outside 2σ of the median are rejected and identified by open diamonds. An order 3 cubic spline is fit to further normalize the flat field on a slice-by-slice level.**

are set in *iraf.nswavelength* to optimize the wavelength solution for the NIFS spectral resolution (*fwidth=2*, *cradius=8* (the maximum distance, in pixels, between a line position and the initial estimate)) as well as to optimize the solution for each grating based on the intensity of the arc lines (see Table 4.2).

The pipeline has predefined input to determine the wavelength solution automatically for the default configurations of the J, H and K gratings. **Though the line list used for the Z grating covers the full band (see Fig. 4.14 and Table 4.2), *iraf.nswavelength* often fails when run automatically on Z-band data. For the Z grating and non-standard wavelength configurations (i.e. the K-long or K-short gratings, or non-standard choices in central wavelength), *iraf.nswavelength* will run interactively and will wait for user input as required. *iraf.nswavelength* does not directly calibrate the data but outputs a series of files in a *database/* directory containing the wavelength solutions to be used later by *iraf.nstransform*. An overview of the *iraf.nswavelength* process for an example K-band arc is shown in Fig. 4.15.**

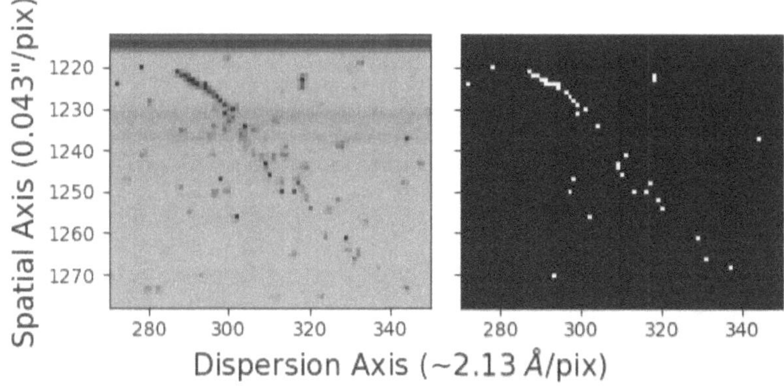

Figure 4.13 **Bad Pixel Mask (BPM) creation via K-band flat frame (*left panel*). Both lamps-on and lamps-off flats are used, however only a lamps-on flat is shown for clarity. Bad pixels are identified via thresholding and are combined to produce a binary BPM (*right panel*) which is used later while reducing the telluric and science data.**

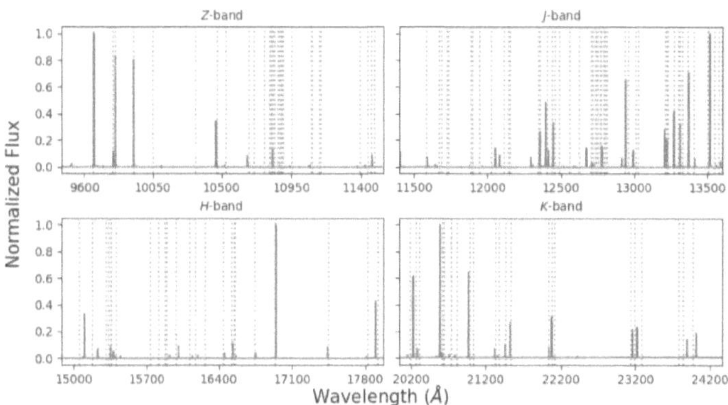

Figure 4.14 **Example of a 1D arc spectrum (blue) and known arc line positions (dashed grey) for each of the four NIFS gratings. The arc spectra shown have been processed with *iraf.nsreduce*, giving only an approximate linear wavelength solution. Arcs are used by *iraf.nswavelength* to compute the wavelength solution (see Fig. 4.15).**

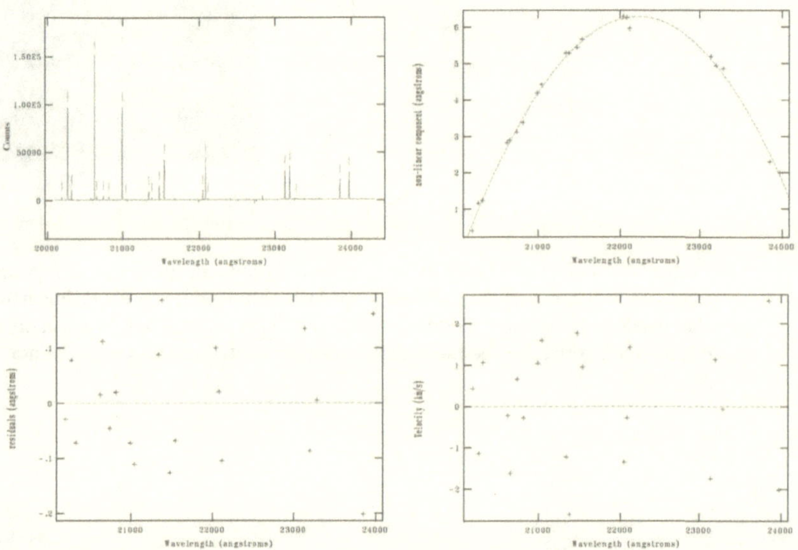

Figure 4.15 **Example output for *iraf.nswavelength* when run on a *K*-band arc. Emission peaks in the arc spectrum are identified and matched to known lines in the line list (*top left panel*). Each identified line is shifted to the rest-frame wavelength and a 4th order chebyshev polynomial is fit to the shifts producing a non-linear wavelength calibration across the full frame (*top right panel*). The wavelength solution is applied to the arc and the positions of the identified line are remeasured. This process is repeated until the line-by-line residuals converge (*bottom left and right panels*; residuals in pixels and km/s, respectively).**

Table 4.2 **Nifty4Gemini wavelength calibration Summary**

Grating	Band Pass (μm)	Calibration Lines	Line List Coverage (μm)	Peak Threshold ADU	N_{lines}	RMS (km/s)
Z	0.94-1.15	Ar, Xe	0.94-1.14	100	33	11.7 ± 2.9*
J	1.15-1.33	Ar	1.16-1.32	100	40	1.6 ± 0.8
H	1.49-1.80	Ar	1.51-1.79	100	24	2.8 ± 1.4
K	1.99-2.40	Ar	2.02-2.39	50	23	5.8 ± 2.9

Note. — Wavelength coverage, line list information, and typical RMS error in the wavelength solution for Nifty4Gemini. *When run automatically (not default); can be minimized with interactive line identification.

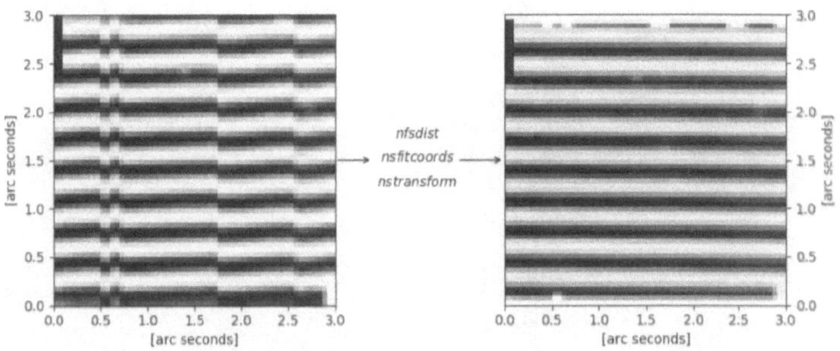

Figure 4.16 **Example of Ronchi frames reconstructed into the sky/telescope plane to demonstrate spatial rectification. The raw frame (*left*) is processed with *iraf.nfsdist*, *iraf.nsfitcoords*, and *iraf.nstransform* (see section 4.2.3) to produce a Ronchi flat that is specially rectified in the sky plane (*right*).**

In the fourth and last task of the NIFS baseline calibration observations reduction, the Ronchi flat frames are respectively being run through *iraf.nfprepare*, combined with *iraf.gemcombine*, and processed with *iraf.nsreduce*, which also applies the flat field.

Finally, *iraf.nfsdist* uses the information in the Ronchi flat calibration image to calibrate the spatial dimension of the NIFS IFU field on the detector. In fact, the measure of the spatial mapping for the images slices with *iraf.nfsdist* is required for accurate alignment of the slices (see Fig. 4.16). The Ronchi flat frame is a dispersed flat field image with a slit-mask in the IFU field so that the illumination on the IFU is in a pattern of 10 different slitlets that are stacked in the y-dimension. Proper alignment of the slitlets across each slice is used for spatial rectification of the NIFS on-sky data. The output of *iraf.nfsdist* is a spatially referenced Ronchi flat with its associated alignment files located in the *database/* directory. The spatial solution determined by *iraf.nfsdist* will be linked to the science and telluric star data by *iraf.nsfitcoords* in later step.

In summary, the outputs of a full NIFS baseline calibration observations reduction are (1) a MDF shift frame, (2) a reduced flat field frame, (3) a bad pixel mask frame, and (4) a wavelength referenced arc frame, and (5) spatial calibration Ronchi flat frame (with an associated database directory, for each of (4) and (5), containing information on the wavelength solution and the spatial distortion correction). All these products are stored in the relevant science target and telluric star observation directories. In the case of standard wavelength configurations observations, the full NIFS baseline calibration observations reduction is run automatically.

Step 3 : Common Steps of the Science Target and Telluric Star Reduction

This stage of the NIFS pipeline is common to both the science target data and the telluric star observations. And done on each set of files located in each science target and telluric star data directories. Step 3 contains eight tasks which are python wrappers around tasks of the Gemini IRAF package.

In this step's first task, all science targets and telluric stars data are processed with *iraf.nfprepare* with the shift image as a reference to apply the MDF shift and add the VAR and DQ extensions. In the second task, sky subtraction is performed. In case of the telluric data, since exposure times are usually short, the sky frames are median-combined with *iraf.gemcombine*. The output combined sky frame is then subtracted from each telluric frame with *iraf.gemarith*. In case of the science data only the closest sky frame in time is subtracted from each science frame with *iraf.gemarith*. In the third task, *iraf.nsreduce* is used to (1) to cut the frames to the size specified in the MDF, (2) to place each IFU slices in separate MEF extensions, (3) to apply the flat field and (4) to apply an approximate wavelength calibrations. **And *iraf.nffixbad* uses the information in the DQ extension, which flagged bad pixels, to correct them by linear interpolation to the nearest pixel not identified as bad along the spatial axis. In the next task, the pipeline computes slice-by-slice 2-dimensional (2D) dispersion and distortion maps with *iraf.nsfitcoords* by fitting a 3rd order chebyshev polynomial to the spatial rectification traces from *iraf.nfsdist* and to the wavelength solution deduced with *iraf.nswavelength*.**

Then *iraf.nstransform* applies the spatial and spectral transformation determined previously by *iraf.nsfitcoords* to produce rectified 3-dimensional (3D) data on a uniformly sampled wavelength scale.

The fifth task uses *iraf.nifcube* to produce a 3D data cube FITS file for each science targets and telluric stars frames. ***iraf.nifcube* takes input from data output by *iraf.nstransform* and by *iraf.nsfitcoords* and converts the 2D data frames into data cubes that have coordinates of x, y, λ and a.square pixel size of 0.05″ on the sky using a 3D Drizzle algortihm. *Nifty4Gemini* calls *iraf.nifcube* with the default input parametres set by the Gemini IRAF Package. A full description of *iraf.nifcube's* functionality, input** parameters and algorithm is given in Sec. C.1. A 1-dimensional (1D) spectrum is extracted from each data cube with *iraf.nfextract* using a circular aperture (the radius is an input parameter in the configuration file), and the pipeline combines all 1D spectra obtained for each observation number for each science target and telluric star respectively. This functionality was added to the pipeline due to a need for a quick and easy way to extract from the datacube 1D spetrum for spatially resolved source located a the center of the NIFS field for some specific type od science. Users should be aware that these 1D spectra are not useful in other cases like (1) extended sources with unknown shape and/or (2) while trying to discover faint emission in the data cubes. In the first case, 'active contours' or other image segmentation methods from computer vision may be useful. In the second problem, see e.g. the DUCHAMP 3D source finder developed for SKA-precursors by Whiting (2012).

Step 4 : Derive and Apply Telluric Absorption Correction to Science 3D data cubes

The first task of the telluric correction is to determine the telluric star spectral type, temperature, the Two Micron All Sky Survey (2MASS; Skrutskie et al. 2006) magnitude by querying SIMBAD (Set of Identifications, Measurements, and Bibliography for Astronomical Data; Wenger et al. 2000) and the observation exposure time from the fits files headers. All these values are then written in a text file and will be used by in the next step to flux calibrate the telluric absorption corrected science data cubes. **The pipeline assumes that a bright, early A-type star was observed to cancel out telluric absorption.** *iraf.telluric* **is used to remove intrinsic absorption lines from the standard star spectrum by shifting and scaling a** $R \sim 5,000$ **spectrum of Vega as a model. The shift and scale values are written out to a text file. While It is recommended to use an early A-type star as telluric standard, if an alternate type of star is chosen, users may specify a custom reference spectrum to be used by emphiraf.telluric. Similarly, the pipeline also provides the option to not remove any intrinsic lines from the telluric star and skip this task.** In that case, the assumption is that either the science will not be affected by those intrinsic lines or the user will provide a telluric star spectrum already corrected if telluric correction will be performed. **After intrinsic absorption is removed, a normalized telluric spectrum is produced by (1) fitting a 3rd order cubic spline to regions not affected by telluric absorption in the telluric star spectrumand (2) by dividing the telluric star spectrum by this continuum fit. Regions free of telluric absorption, in each grating, have been identified by eye and are incorporated into the pipeline, however users are encouraged**

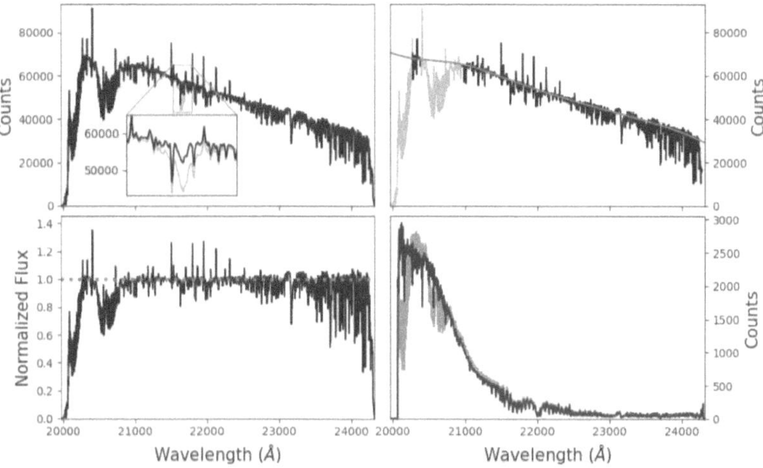

Figure 4.17 **Example of the telluric correction procedure for a sample K-band standard star and science frame.** *Top left:* **Extracted 1D telluric standard spectrum from a reduced telluric standard data cube. The spectrum prior to H-line removal is shown in gray while the H-line corrected spectrum is shown in black.** *Top right:* **Continuum fit with a 3rd order cubic spline (red) to the extracted spectrum. Regions heavily affected by telluric absorption (grey) were masked and therefore not included in the continuum fit.** *Bottom left:* **Continuum normalized telluric spectrum. This normalized spectrum is scaled and shifted via *iraf.telluric* to best remove telluric absorption lines from the science frames.** *Bottom right:* **Extracted uncorrected 1D science spectrum (gray) and telluric corrected science spectrum. Telluric corrections are performed on the individual science exposures before merging the cubes.**

to fit these regions interactively for higher fidelity telluric correction.

In the final task, *iraf.telluric* is used to remove the telluric absorption from the 3D science data cube by shifting and scaling the normalized telluric spectrum and dividing it from the science cube. While *iraf.telluric* performs an iterative fit to optimize the telluric absorption correction, this process is fundamentally limited as only a single shift and scale value is ultimately used. A proper treatment of telluric removal, by assessing each telluric line individually, is encouraged for users who require high precision telluric correction.

Step 5 : Flux calibrate the Telluric Absorption Corrected Science 3D data cubes.

Flux calibration for NIFS spectra is performed as in version 1.9 of the *XDGNIRS* pipeline (Mason et al., 2015). A blackbody spectrum is produced using the known temperature of the standard star and is then scaled to the standard star's 2MASS magnitude (Skrutskie et al., 2006). This scale is then applied to all science cubes. Assuming A-type stars have been used for the telluric standard, this assumption is reasonable as the nearly featureless spectra of A-type stars in the IR closely match the continuum spectrum of a blackbody of the appropriate temperature (Pecaut & Mamajek, 2013).[12] Mason et al. (2015) assessed the accuracy of this calibration by observing multiple standard stars on a single night and using one star to flux calibrate the other(s). Flux errors of 5-40% were observed, with seeing variations within the 0.3" GNIRS slit contributing the largest source of error.

Step 6 : Merge all Science 3D data cubes of the same target

The Gemini IRAF Package does not provide any task to merge the 3D science cubes and this process can be challenging if there are many data cubes obtained on different nights. Thus, *Nifty4Gemini*'s NIFS pipeline provides an efficient and flexible process to merge all science 3D data cubes of the same target into a final cube. This step is composed of six tasks, which can be used to combine the three different types of 3D science data cubes produced by the pipeline : (1) data cubes that are produced by *iraf.nifcube* at the end of step 3 of the data reduction process, (2) data cubes which have been corrected from telluric absorptions and (3) data cubes which have been corrected from telluric absorptions and flux calibrated. The users can chose to do all of these tasks. **It is improtant to note that none of these tasks changes the data cubes pixels size produced by *iraf.nifcube*, and the only combining operations performed on the data cubes pixels are those done by *iraf.imcombine* only and the choices are *average, median* and *sum*.**

In addition, the pipeline has currently three ways to combine 3D data cubes. In the first case, the pipeline uses the Gemini coordinates system for telescope offsets defined as *(p,q)*, which are in arcsec, and read the p and q values from the MEF file headers. The pipeline goes through all the directories which contains science 3D data cubes, and makes a list of the files names. After it copies each of the files in sub-directories under a new created top directory called *Merged*. The sub-directories's names are the observation ID.

---

[12] An interactive approach to the flux calibration is encouraged if the observer chooses to use a telluric standard of an alternative spectral class.

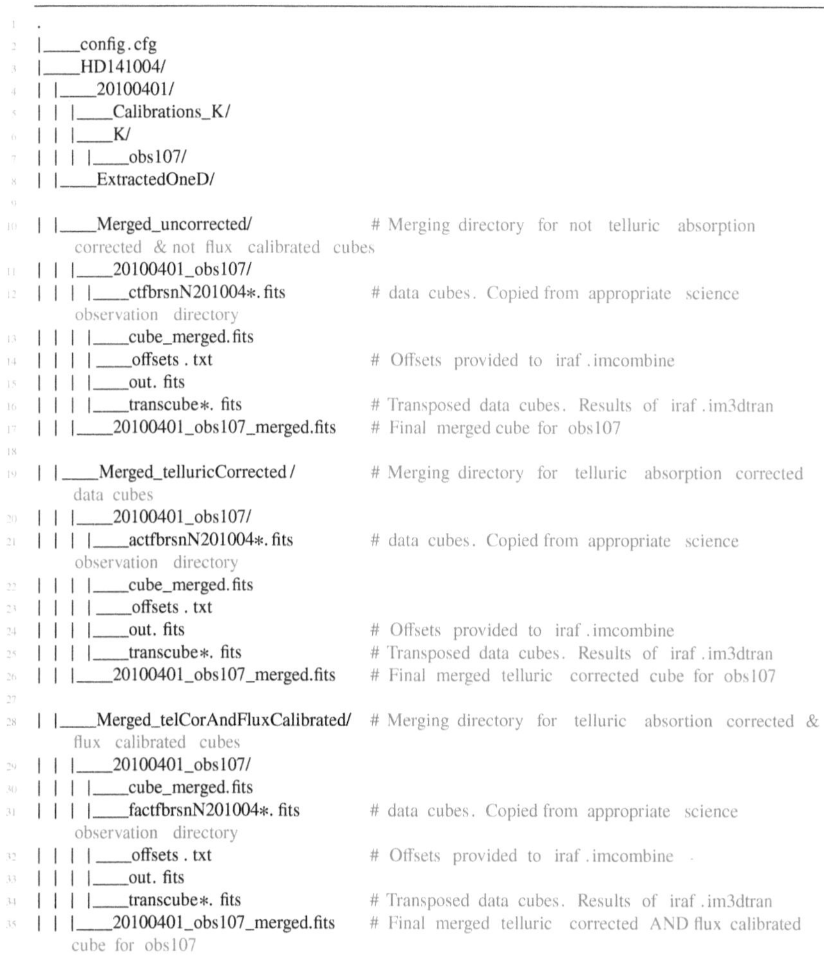

```
.
|____config.cfg
|____HD141004/
|  |____20100401/
|  |  |____Calibrations_K/
|  |  |____K/
|  |  |  |____obs107/
|  |____ExtractedOneD/

|  |____Merged_uncorrected/              # Merging directory for not  telluric  absorption
         corrected & not flux  calibrated  cubes
|  |  |____20100401_obs107/
|  |  |  |____ctfbrsnN201004*. fits       # data cubes. Copied from appropriate  science
         observation  directory
|  |  |  |____cube_merged.fits
|  |  |  |____offsets . txt               # Offsets  provided  to  iraf .imcombine
|  |  |  |____out. fits
|  |  |  |____transcube*. fits            # Transposed data cubes.  Results  of  iraf .im3dtran
|  |  |____20100401_obs107_merged.fits    # Final merged cube  for  obs107

|  |____Merged_telluricCorrected /        # Merging directory for  telluric  absorption  corrected
         data cubes
|  |  |____20100401_obs107/
|  |  |  |____actfbrsnN201004*. fits       # data  cubes.  Copied from appropriate  science
         observation  directory
|  |  |  |____cube_merged.fits
|  |  |  |____offsets . txt
|  |  |  |____out. fits                    # Offsets  provided  to  iraf .imcombine
|  |  |  |____transcube*. fits            # Transposed data cubes.  Results  of  iraf .im3dtran
|  |  |____20100401_obs107_merged.fits    # Final merged telluric   corrected cube  for  obs107

|  |____Merged_telCorAndFluxCalibrated/   # Merging directory for  telluric  absortion  corrected  &
         flux  calibrated  cubes
|  |  |____20100401_obs107/
|  |  |  |____cube_merged.fits
|  |  |  |____factfbrsnN201004*. fits      # data  cubes.  Copied from appropriate  science
         observation  directory
|  |  |  |____offsets . txt               # Offsets  provided  to  iraf .imcombine
|  |  |  |____out. fits
|  |  |  |____transcube*. fits            # Transposed data cubes.  Results  of  iraf .im3dtran
|  |  |____20100401_obs107_merged.fits    # Final merged telluric   corrected AND flux  calibrated
         cube  for  obs107
```

Figure 4.18 Example of the merging directories after running the pipeline to get merged 3D science data cubes for a single observation (number 107) taken on a single night (Apr. 04, 2010).

```
|____Merged_uncorrected        # Merging directory for not telluric absorption corrected & not
     flux calibrated cubes
| |____.DS_Store
| |____20130527_obs28_merged.fits  # Final merged cube for observation number 28
| |____20130530_obs36_merged.fits  # Final merged cube for observation number 36
| |____20130530_obs55_merged.fits  # Final merged cube for observation number 55
| |____20130531_obs36_merged.fits  # Final merged cube for observation number 36 on 2013, May
     31
| |____20130621_obs36_merged.fits  # Final merged cube for observation number 36 on 2013, June
     21
| |____20130622_obs44_merged.fits  # Final merged cube for observation number 44
| |____20130624_obs75_merged.fits  # Final merged cube for observation number 75
| |____20130626_obs83_merged.fits  # Final merged cube for observation number 83
| |____temp_mergedH.fits
| |____TOTAL_mergedH.fits # Final Total S/N not telluric absorption corrected & not flux
     calibrated cube
| |____waveoffsetsH.txt # Offsets file provided to iraf .imcombine
|____Merged_telluricCorrected # Merging directory for telluric corrected data cubes
| |____.DS_Store
| |____20130527_obs28_merged.fits  # Final merged cube for observation number 28
| |____20130530_obs36_merged.fits  # Final merged cube for observation number 36
| |____20130530_obs55_merged.fits  # Final merged cube for observation number 55
| |____20130531_obs36_merged.fits  # Final merged cube for observation number 36 on 2013, May
     31
| |____20130621_obs36_merged.fits  # Final merged cube for observation number 36 on 2013, June
     21
| |____20130622_obs44_merged.fits  # Final merged cube for observation number 44
| |____20130624_obs75_merged.fits  # Final merged cube for observation number 75
| |____20130626_obs83_merged.fits  # Final merged cube for observation number 83
| |____temp_mergedH.fits
| |____TOTAL_mergedH.fits # Final Total S/N telluric corrected cube
| |____waveoffsetsH.txt # Offsets file provided to iraf .imcombine
|____Merged_telCorAndFluxCalibrated # Merging directory for telluric absorption corrected &
     flux calibrated cubes
| |____.DS_Store
| |____20130527_obs28_merged.fits  # Final merged cube for observation number 28
| |____20130530_obs36_merged.fits  # Final merged cube for observation number 36
| |____20130530_obs55_merged.fits  # Final merged cube for observation number 55
| |____20130531_obs36_merged.fits  # Final merged cube for observation number 36 on 2013, May
     31
| |____20130621_obs36_merged.fits  # Final merged cube for observation number 36 on 2013, June
     21
| |____20130622_obs44_merged.fits  # Final merged cube for observation number 44
| |____20130624_obs75_merged.fits  # Final merged cube for observation number 75
| |____20130626_obs83_merged.fits  # Final merged cube for observation number 83
| |____temp_mergedH.fits
| |____TOTAL_mergedH.fits # Final Total S/N telluric corrected & flux calibrated cube
| |____waveoffsetsH.txt    # Offsets file provided to iraf .imcombine
|____Nifty.log         # Log file for the entire data reduction process with input and output for
     the PyRAF tasks.
```

Figure 4.19 Example of a directory structure showing all merged 3D science cubes as well as the total S/N merged cube for the three differents types of science data cube produced: not telluric absorption corrected and not flux calibrated, only telluric absorption corrected and telluric absorption corrected and flux calibrated.

For each sub-directory, the *(p, q)* values from the first cube in the list is used as the reference. Thus, the pipeline computes the x and y offsets from these references values for each of the remaining cube from the list. Those offsets are converted into pixels and stored in files called *offsets.txt*. The references *(p, q)* values are recorded as *(x=0, y=0)* in pixels size in the offsets.txt file. The third axis, corresponding to the wavelength, is also recorded as *0* for all cubes. After the pipeline uses *iraf.imcombine* to apply the shifts in the *offsets.txt* to all data cubes before to combine them in a single 3D data cubes. However before this step the pipeline, by default, swaps the lambda and y-axis using *iraf.im3dtran* (the "offsets.txt" format is *(x, λ, y))*. We found that if this is done *iraf.imcombine* runs 50 times faster, otherwise it can takes 25 minutes to shift and merge cubes. We have verified that cubes produced with and without using *iraf.im3dtran* are identical. The users can decide or not to use *iraf.im3dtran*. The pipeline will swap back the lambda and y-axis after the merging. The second way to combine the 3D cubes is to provide an already completed "offsets.txt" files. In this case, the pipeline reads in the offsets file and uses it with *iraf.imcombine* as described before. As the third way, the user can provide already shifted 3D data cubes. In that case, while scanning each of the directories with science 3D data cubes, if the pipeline find files names which start with "shift" it will directly combine them with *iraf.imcombine*. Figure 2 shows an example of the merging directories after running the pipeline to get merged 3D science data cubes for the three types of science data cubes. In this example, there is only one single observation (e.g. observation 107), and then the pipeline produced merged cubes for telluric absorption corrected and flux calibrated cubes, telluric absorption corrected only cubes and cubes with no telluric absorption correction and no flux calibration done.

Once a merged cube has been produced for each sub-directory, a list is created of all merged cubes for each day observations were taken with the same grating configuration. The pipeline then selects the first cube in the list as the reference *(x=0, y=0)* and compute the x and y offsets from these references values for each of the remaining cube from the list. It also computes any wavelength offsets from the other cubes and include them in the text file containing the offsets (for example *waveoffsetsK.txt*). Finally, the pipeline uses *iraf.imcombine* to create the final 3D data cube, by merging all those cubes. This last merged data cube should have the full S/N from all science exposures of a given NIFS grating configuration for the same science target. Figure 3, shows an example of the merging directories after running the pipeline to get merged 3D science data cubes for each days observations were taken, and after the pipeline produced a final full S/N 3D science cube with the same grating configuration for a single science target.

Fig. 4.20 shows an example of the merging process for a non-sidereal target, Saturn's moon Titan. In this case, because Titan's coordinates are not the same between merged cubes, it is not possible to just use *(p, q)* offsets to combine the merged science 3D data cubes from different nights into a final total S/N science data cube. It is a good example when the user should update manually the *waveoffsetsK.txt* file. To reduce the Titan's NIFS data : (1) the pipeline is ran once, using the configuration file from Figure 5, and produced offsets files (e.g. *offsets.txt*) for each night of observation, as well as the *waveoffsetsK.txt* file), merged cubes for each nights as well as the final full S/N cube; (2) only the wavelength offsets should be correct in the *waveoffsetsK.txt* file, thus the correct x and y offsets are computed manually for each merged cubes on different nights and the *waveoffsetsK.txt* file is updated; (3) the pipeline is ran one more time to perform only the last merging task, by using the updated *waveoffsetsK.txt* file with *iraf.imcombine* and to create the correct full S/N 3D science cube.

Tutorials and Examples

The online documentation contains various examples in order to illustrate the flexibility of *Nifty4Gemini*, and its associated NIFS PyRAF/python based pipeline. The online documentation contains multiple examples of config.cfg configuration files for science programs which can be accessed from the Gemini public archive.[13] Figure 5 shows the config.cfg configuration file used to reduce the NIFS raw data on Titan, obtained by program GN-2014A-Q-85. There is also an example of the config.cfg configuration file for which the sorting of the raw data is done manually and the user specifies the various paths to the different type of data. This especially relevant when data for a taken on multiple days for the same observation ID in the Gemini Observing Tool (OT). The config.cfg configuration file is also provided for some tutorial data that can be downloaded directly from the NIFS web page. [14] In addition to example of config.cfg configuration files, Examples of directory structures are provided after running each of the step of the data reduction process. [15] The purpose is to help the user to identify quickly the important outputs of each steps and to be able to assess if something is missing. Finally the online documentation contains various tutorials. [16] The purpose is to teach the users how to run some of the tasks interactively (e. g. intrinsic star absorptions removal), as well as how to incorporate their own reduced

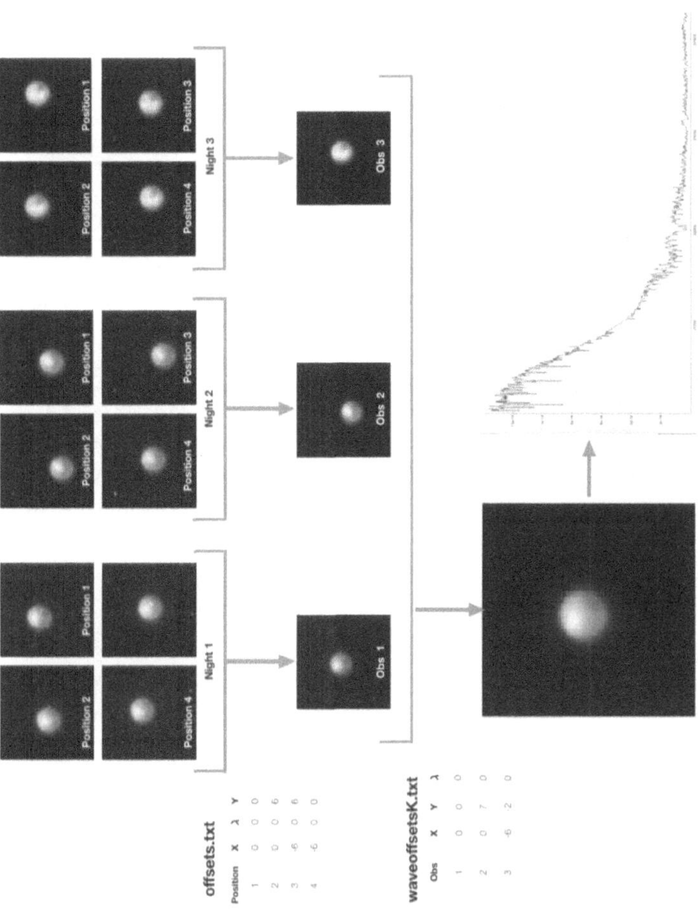

Figure 4.20 Example of merging sequence for NIFS data of Titan (program GN-2014A-Q-85; obtained in Apr./May 2014). For each observation number, taken on a different night, a merged cube is produced by using the $(x,\ \lambda,\ y)$ offsets, in pixels, obtained from the four $(p,\ q)$ telescope positions. Since Titan is non-sideral target, the RA and Dec are not the same for each of the three merged cubes. Therefore a total S/N data cube is made after provided a manually updated *waveoffsetsK.txt* file and running the last merging step of the pipeline. A 1-D spectrum has been extracted from the final total S/N data cube with an aperture of 0.9 arcsec using QfitsView.

data to the pipeline (e. g. provide own continuum-normalized telluric spectrum for telluric correction).

4.2.4   Conclusion and Future Outlook

In this chapter, I presented an overview of the current state of *Nifty4Gemini*, a python package which includes a data processing pipeline for NIFS observations. The NIFS pipeline incorporates all steps needed to produce a 3D data cube with the total S/N from all the acquired science exposures for a given NIFS grating configurations of the same science target. *Nifty4Gemini* is under continuous development. The online documentation lists the current known issues. Some of them are permanent limitations (e. g. IRAF task parameters must be 99 characters or less) while others can be addressed by further work (current telluric correction not built to be run automatically). There are plans to address the latter in future release. In addition, future small and large possible extensions of *Nifty4Gemini* and its associated PyRAF/python based pipelines are described under the "Future work" section of the online documentation. For instance, one extension is to have *Nifty4Gemini*'s NIFS pipeline run automatically and internally inside the Gemini Observatory to reduce priority data and have them made available to the PI of NIFS programs the next day after the observation were taken. Another planned major extension is to incorporate to *Nifty4Gemini*, the *XDGNIRS* python pipeline for GNIRS cross-disperser mode. Finally, any maintenance of the NIFS instrument that can affect the data reduction or changes of the Gemini IRAF package and AstroConda should trigger a new update of the package.

4.3   Data Products for Gemini GRACES Spectra

This section provides a short overview of a python-based data reduction suite I produced to improve the quality and usefulness of the OPERA (Martioli et al., 2012) reduced stellar spectra that are available on the Gemini Data Archive. The spectra used in Chapters 2 and 3 were processed with this pipeline, and the chemical abundances in Chapter 3 are data products of the final stages of this method. Though still very much a "by hand" method, the creation of these data reduction and synthetic spectrum synthesis tools has dramatically increased the efficiency and reproducibility of my chemical abundance analysis. Future Gemini instruments like GHOST may also benefit from the data rectification and continuum normalization schedule implemented here. I plan on working these tools into a shareable, easy-to-use, open-source package which can be utilized by future stellar spectroscopists.

```
# Nifty4Gemini configuration file.
#
# Each section lists parameters required by a pipeline step.

niftyVersion = '1.0.0'
manualMode = False   # if True user need to press the enter key for execution of the next task/step.
over = False             # overwrite of output, if False pipeline skips already executed steps.
extractionXC = 15.0    # X coordinate for 1-D spectrum extraction
extractionYC = 33.0   # Y coordinate for 1-D spectrum extraction
extractionRadius = 2.5 # radius in arcsec of circular aperture for 1-D spectrum extraction
scienceOneDExtraction = True  # if True 1-D spectrum extraction will be performed on science cubes.
scienceDirectoryList = []        # directory where science data are.
telluricDirectoryList = []        # directory where telluric star data are.
calibrationDirectoryList = []     # directory where calibration data are.

[nifsPipelineConfig]
sort = True              # if True perform Step 1 : Getting and Sorting the NIFS raw data
calibrationReduction = True # if True perform Step 2 : Reducing the NIFS raw calibration data
telluricReduction = True     # if True perform Step 3 : Telluric Star Reduction
scienceReduction = True  # if True perform Step 3 : Science Reduction
telluricCorrection = True    # if True perform Step 4 : Derive and Apply Telluric Absorption Correction to Science 3D data cubes
fluxCalibration = True    # if True perform Step 5 : Flux calibrate Telluric Absortion Correction Science 3D data cubes
merge = True             # if True perform Step 6 : Merge all Science 3D data cubes of the same target
telluricCorrectionMethod = 'gnirs' #Telluric Correction method
fluxCalibrationMethod = 'gnirs'   #Flux calibration method
mergeMethod = ' '         #Cubes' Merging method

[sortConfig]
rawPath = ' '          # Directory of the raw data
program = 'GN-2014A-Q-85' # Program ID if raw data needs to be downloaded from the Gemini Archive
proprietaryCookie = ' '       # not implemented yet.
skyThreshold = 2.0 #used to differentiate between sky frame and science from the telescope offsets defined as \emph{(p,q)}
sortTellurics = True   #If True will sort the telluric star data and match them to the science data.
telluricTimeThreshold = 7200   # time limit in second between a science frame and a telluric star frame

[calibrationReductionConfig]
baselineCalibrationStart = 1 # First task for the calibration data reduction
baselineCalibrationStop = 4 # Last task for the calibration data reduction

[telluricReductionConfig]
telStart = 1        #First task for the telluric star reduction
telStop = 5 # Last task for the telluric star reduction
telluricSkySubtraction = True #If True perform sky substraction on the Telluric data

[telluricCorrectionConfig]
telluricCorrectionStart = 1  #First task for the telluric   correction
telluricCorrectionStop = 9 # Last task for the telluric correction
hLineMethod = 'vega' #H lines removal method
hLineInter = False   #if False turn off interactive mode
continuumInter = False #if False turn off interactive mode
telluricInter = False   #if False turn off interactive mode
tempInter = False #if False turn off interactive mode
standardStarSpecTemperature = ' ' #Blackbody temperature for telluric star spectral type.
standardStarMagnitude = ' ' #magnitude of the telluric star
standardStarRA = ' '  #RA of the telluric star
standardStarDec = ' ' #Dec of the telluric star
standardStarBand = ' ' #Filter for the magnitude

[fluxCalbrationConfig]
fluxCalibrationStart = 1  #First task for the flux calibration
fluxCalibrationStop = 6 #Last task for the flux calibration

[mergeConfig]
mergeStart = 1 #First task for the merging of the science 3D cubes
mergeStop = 6 #Last task for the merging of the science 3D cubes
mergeType = 'median' #combining method of the science cubes
use_pq_offsets = True #If True uses the Gemini coordinates system for telescope offsets defined as \emph{(p,q)}
im3dtran = True #If True swaps the lambda and y axis using iraf.im3dtran before combining cubes.

# Good luck with your Science!
```

Figure 4.21 Example of the config.cfg configuration file used to reduce the NIFS raw data on Titan, obtained by program GN-2014A-Q-85.

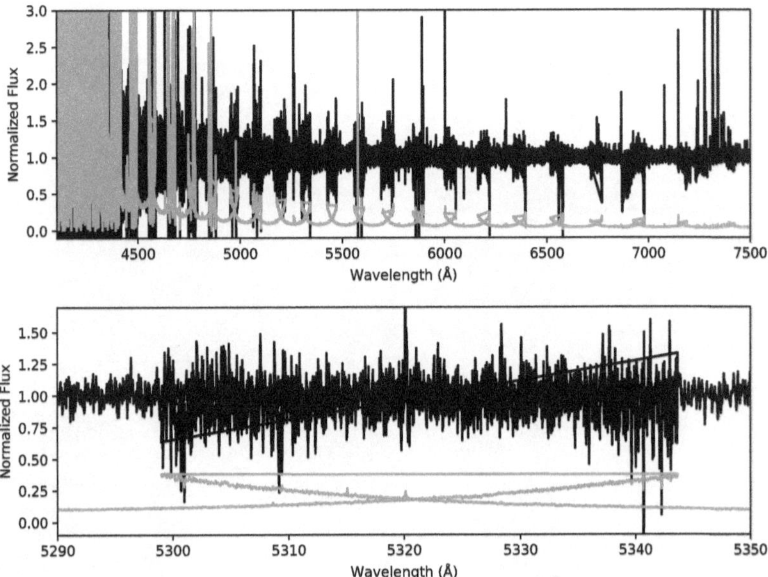

Figure 4.22 Example of an OPERA reduced GRACES spectrum for a star in the *Pristine-GRACES* sample. The top panel shows the individual visit spectrum (black) over wide spectral window, highlighting the poor SNR in the blue and in the regions where the echelle orders overlap. The error spectrum (i.e. the uncertainty on the flux) provided by OPERA is shown in orange. The bottom panel is a zoom in of this spectrum demonstrating the increased noise at the edges of each echelle order.

The pipeline starts with the individual visit spectra for each star. Since the spectra are produced by an echelle spectrograph, the OPERA reduced 1D spectra from ESPaDOnS/-GRACES are stitched together from each extracted echelle order. For two subsequent orders n and $n + 1$, there is often an overlap between the two orders in the wavelength coverage which can hide weak spectral lines (like most of those seen in the spectra of EMP stars) This is shown in Fig. 4.22.

To boost the SNR in each overlapping region, the two overlapping regions are combined via a weighted average using the error spectrum provided by OPERA is used as weights. The construction of the 1D OPERA spectrum, as a data array, "jumps" backwards in the wavelength dimension (shown by the straight lines in Fig. 4.22); the pipeline identifies

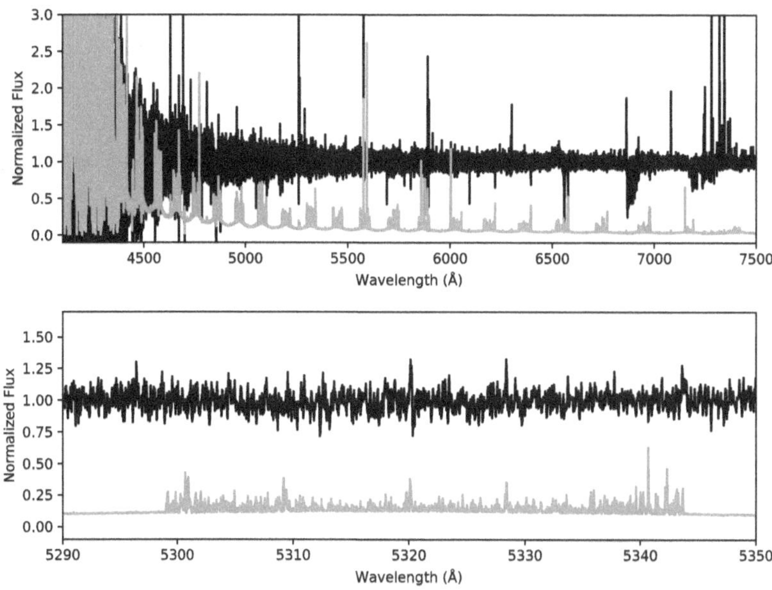

Figure 4.23 Same as Fig. 4.22 but where the overlapping regions between echelle orders are combined via a weighted average. The SNR in these overlapping regions is improved substantially.

where these jumps occur, trims the spectrum accordingly, performs the weighted average, and then reassembles the full spectrum. After this first step, the SNR of the individual visit spectrum is dramatically improved (e.g. Fig. 4.23).

The following steps in the reduction are standard in the rectification of stellar spectra:

- All visits for a given star are examined for RV variations. A "first-pass" RV correction for each visit is performed via a cross-correlation with a high SNR comparison spectrum of metal-poor star HD 122563. If the RV's measured between each visit are equal, to within the precision of the wavelength solution (usually ~0.5km/s), then the star is not flagged for potential binarity. Due to the short cadence of the observations in these programs, there has been no observed evidence of binary systems based on this criteria.

- All visits for a given star are coadded via a weighted mean using the visit error spectra

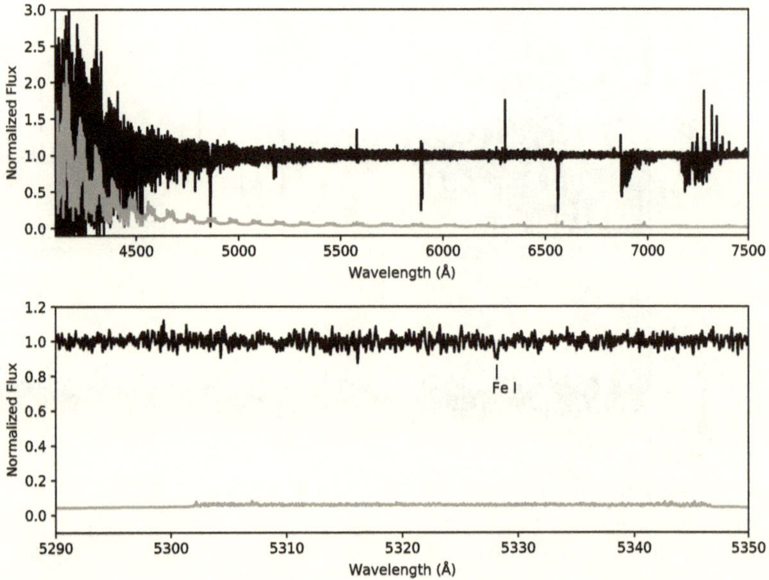

Figure 4.24 Example of a fully rectified GRACES spectrum for the same star shown in Figs. 4.22 and 4.23. Only after the full rectification and coaddition of the eight individual visit spectra can the Fe I line at 5328.039 Å be seen.

to provide the weights.

- The higher SNR coadded spectrum is RV corrected as described above.

- The RV corrected and coadded spectrum is re-normalized using an asymmetric sigma-clipping routine.

These final, flat spectra are used in all future analysis. An example of the final spectrum for the star shown in Figs. 4.22 and 4.23 is given in Figure 4.24. The Fe I line at 5328.039Å, one of the most diagnostic lines used to determine [Fe/H] for the EMP stars in the GRACES sample, is only visible after the combination of 8 fully rectified visit spectra (though the signal is still below a 3σ detection).

The pipeline up to this stage provides a useful data product that stellar spectroscopists may then use in their own chemical abundance analysis procedures. The remaining steps in the suite are effectively a python wrapper for a synthetic spectrum synthesis using MOOG:

- A MARCS model atmosphere is created using the predetermined Bayesian inferred stellar parameters T_{eff} and $\log g$, and [Fe/H] from MRS/FERRE analysis. The user could also specify their own stellar parameters or a file to pull from. A cube of eight model atmospheres, which encapsulate these specified stellar parameters, is found within a larger, pre-generated grid of MARCS models. The model atmosphere for these specified parameters is created via an interpolation of the cube.

- Using a line list of ~ 300 known isolated (non-blended) Fe I and II lines, the observed spectrum is automatically examined to identify Fe lines from which an abundance measurement can be made. The user can specify the upper-limit criteria or choose to run the analysis with upper-limits included. A truncated Fe line list is then created.

- All configuration files for MOOG are automatically populated with references to the observed spectrum, model atmosphere, line lists, and abundance grids to iterate over.

- In the first stage of iterations, the Fe abundance is refined from the original MRS estimate. Each of the vetted lines from the line list is synthesized at a specified abundance. A least-squares fit between the synthetic spectrum and the observed spectrum is computed over a small window (~ 10Å) around each identified Fe line. The Fe abundance of the synthetic spectrum is changed in MOOG, a new spectrum is synthesized, and the fit for each line is performed again.

- This process is repeated over a large range of metallicities ($-7 \lesssim$ [Fe/H] $\lesssim 0$) with a step size of 0.05 dex until a best fit Fe abundance is determined for each Fe line.

- The uncertainty on the Fe measurement for each line is determined by the SNR of the observed spectrum around each line. Assuming the continuum flux should be 1.0 in normalized flux units, the scatter of the observed spectrum around 1.0, in regions free of absorption lines, can be used to define an error in the continuum placement (e.g. 1.00 ± 0.05, in normalized flux units). Synthetic spectra of varying abundances, for an individual line, are compared to the best fit synthetic model for that line until the difference in peak absorption between the two syntheses is equal to the error in continuum placement. This represents a 1σ error in the abundance measurement.

- The abundances determined from all Fe lines are averaged with a weighted average, and this updated [Fe/H] is used to create a new model atmosphere. The above steps are repeated until the metallicity converges (usually only after two iterations).

- Once a metallicity has been finalized and a final model atmosphere is produced, the line-by-line fitting process is repeated using a series of line lists for Li, Na, Mg, K, Ca, Sc, Ti, Cr, Mn, Ni, Cu, Zn, Y, Zr, Ba, La, Nd, and Eu. If an element is known to exist in a blended region or if it is subject to hyper-fine structure splitting (HFS), rather than synthesizing the individual line as was done for Fe, a region of ± 5Å around the line of interest is synthesized in MOOG using a line list containing relevant lines for blends and/or atomic information to account for HFS splitting. These lists can be easily generated and this process can be expanded to other elements and other spectral regimes. Only the α-elements are added into the final model atmosphere, as the atmosphere structure is unlikely to change substantially due to the presence of the other heavier elements.

- Finally, a systematic abundance analysis is performed by adjusting the model atmosphere by the errors on T_{eff}, $\log g$, and [Fe/H] (each error is examined independently) and new best-fit abundances are determined for all species.

The output data products produced by this method are familiar to MOOG users and can be easily incorporated into one's own existing scripts to produce chemical abundance plots and tables. It is my hope that these tools can be used by future graduate students (and other researchers) to spare them from the nightmares encountered when running MOOG by hand.

Chapter 5

Ongoing Projects and Conclusions

5.1 The *Pristine* Survey: Gemini-GRACES chemo-dynamical study of newly discovered very metal-poor stars in the Galactic Bulge

In Chapter 3, I presented the chemo-dynamical study of very metal-poor stars found within the *Pristine* survey. The high-resolution spectra for these stars were collected as part of a Gemini/GRACES Large and Long Program (LLP). These spectra were necessary to uncover chemically unique stars and they provided precision RV's which are needed to compute dynamics. A majority of these stars were found to be members of the Galacitc halo, though a few stars on planar like orbits were also found. Previously discussed in sections 1.1.3 and 3.1, the Galactic halo is not the only location in the Galaxy that EMP stars are expected to be found. Motivated by the knowledge that the oldest metal-poor stars in the Galaxy are thought to exist within the Galactic bulge (Tumlinson, 2010; Starkenburg et al., 2017b; Lucey et al., 2019), a subset of the time awarded in the LLP was allocated to following-up promising metal-poor bulge members found within the *Pristine* survey.

The Pristine Inner Galaxy Survey (PIGS) (Arentsen et al., 2020b,a) is a new photometric and spectroscopic program to find metal-poor stars in the highly crowded and inhomogeneously extincted Galactic bulge. Arentsen et al. (2020a) successfully followed-up $\sim$ 8000 metal-poor candidates in the magnitude range $14 < V < 17$ with low resolution spectroscopy using AAOmega+2dF at the Anglo-Australian Telescope, uncovering 1300 VMP stars - the largest sample of VMP stars in the inner Galaxy to date. Detailed chemical analyses of those metal-poor stars are needed to determine whether the star formed in situ, has fallen in, or is just passing through the bulge along our line of sight (Lucey et al., 2020). *Gaia* DR2 (Gaia Collaboration et al., 2018) proper motions can play a critical role in identifying stars that are members of the bulge from their orbits; however, accurate distances to the individual objects are necessary and often lacking. Only a few targets have sufficiently accurate parallaxes towards the Galactic bulge for precision distance estimates. In this section, I present an ongoing follow-up project for four of the most metal-poor candidates found by Arentsen et al. (2020b,a) which were observed as part of the Gemini/GRACES LLP. The GRACES spectra for these four stars were processed and rectified in an identical fashion described in Chapter 3.

5.1.1 Stellar parameters and preliminary chemical abundances

Both ULySS+MILES (Koleva et al., 2009; Prugniel et al., 2011; Sharma et al., 2016) and FERRE Aguado et al. (2019b) are data analysis pipelines that use model atmospheres and synthetic spectra to determine stellar parameters (T_{eff}, log g, [Fe/H], and [C/Fe]) for individual stars. Both pipelines were used to calculate stellar parameters from the low resolution AAOmega+2df spectra, but the ULySS stellar parameters were initially adopted. Known degeneracies between FERRE log g and [C/Fe] were dissuading, and the large uncertainties in the *Gaia* data were prohibitive in determining stellar parameters via the Bayesian inference method.

Preliminary Fe abundances have been determined from a 1D-LTE MARCS model atmospheres analysis, using the MOOG spectrum synthesis program in its' equivalent width(EW) abundance finding routine. EWs are measured for clean spectral lines selected from the detailed synthetic analyses in Chapter 3. Spectrocopic checks on T_{eff} and log g were performed, examining the slope of A(FeI) vs. excitation potential (χ in eV) and the FeI to FeII ionization equilibrium. The spectroscopic stellar parameters determined for three of the four stars are in agreement with the ULySS parameters, however P182129, the most metal-poor star in the sample at [Fe/H]$_{EW} \sim -3.6$, showed significant deviations from the previously calculated parameters.

Both the ULySS and FERRE models for P182129 give temperatures and gravities which are too high. The source of this issue is predicted to arise from the blue sensitive AAOmega spectra and the high degree of dust reddening towards the Galactic bulge. As dust absorbs blue light and re-emits it at redder wavelengths, the observed continuum in the blue is dampened. Consequently, absorption lines will appear weaker, an effect also produced in hotter atmospheres. Since Fe lines measured in the GRACES spectrum are at redder

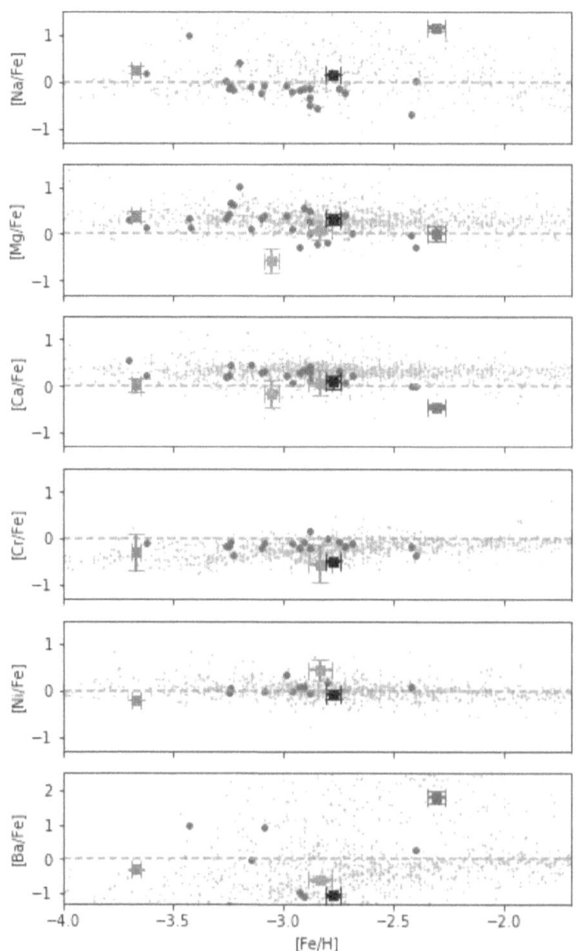

Figure 5.1 Preliminary LTE chemical abundances for Na, Mg, Ca, Cr, Ni, and Ba for the GRACES bulge sample. The four stars studied are in coloured points. Dark grey circles are from Chapter 3 and background stars (light grey dots) are from (Venn et al., 2004; Aoki et al., 2013; Yong et al., 2013; Roederer et al., 2014; Frebel et al., 2014)

wavelengths, I expect dust extinction is less of an issue and the abundances measured from these lines are more accurate than those determined by the AAOmega spectra (not to mention the much higher resolution of GRACES; $R \sim 1300$ vs. $R \sim 60,000$). This exploration of dust extinction towards the bulge and its' effect on blue sensitive spectra is an ongoing project. The spectroscopic T_{eff} and $\log g$ for P182129 are currently used in the rest of the analysis.

Preliminary Na, Mg, Ca, Cr, Ni, and Ba abundances have also been determined via EWs and are presented in Figure 5.1. I anticipate these abundances will vary once a full synthetic spectrum synthesis, as in Chapter 3 is performed, so I will reserve my commentary on the chemical abundance profiles of these stars until then.

5.1.2 Dynamics of the bulge candidates

It is important to distinguish stars that may have formed in situ in the bulge versus those that are halo or disk stars just passing through our line of sight (Lucey et al., 2020). The best way to do this is from their Gaia proper motions and an estimate of their orbits. Following (Howes et al., 2016, 2015, 2014), I consider a star that is within 5 kpc of the Galactic bulge to be an in situ member.

Determining distances to these four stars is non-trivial. The large errors in parallax ($> 20\%$) mean that a simple $d = 1/\varpi$ is ill-advised (Bailer-Jones, 2015) and the distances inferred from the Bayesian analysis are currently biased to disk and halo stars. To study the dynamics of these stars, orbits are calculated over a range of assumed distances that run through the bulge and along our line of sight. Orbits are calculated assuming the distances are fixed at 4, 6, 8, and 10 kpc, a range that bounds the parallax distance measurements for this sample and would place the star within the bulge, from our side of the Galaxy to slightly past the Galactic Centre. Folding in the precision radial velocities from our high resolution spectra, and proper motions from the *Gaia* DR2 database, dynamics were calculated as in Chapters 2 and 3 using Galpy (Bovy, 2015).

In Fig.5.2, the range of orbits for each star are shown at each fixed distance. There is no distance for two stars (P180503 or P182129) that would place them in the bulge; in both cases r_{apo} is always greater than ~ 8 kpc and z_{max} is always greater than ~ 5 kpc. For the other two stars (P183335 and P184700) a range of distances are found that could place them in the bulge $r_{apo} \leq 5$ kpc - and, in fact, on planar orbits $z_{max} \leq 2$ kpc.

When examining the action parameters, some of the larger fixed distances can put stars on retrograde orbits, and I note that the largest distance assumed for P180503 could put it on

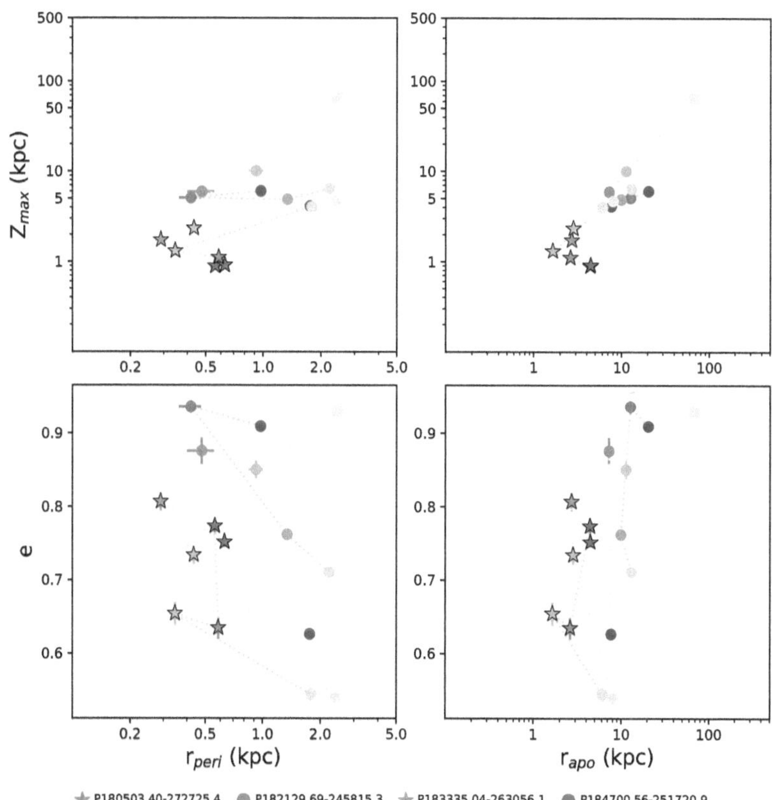

Figure 5.2 Orbital elements calculated with Galpy for the GRACES Bulge sample. The left column shows the max perpendicular distance from the Galactic plane (z_{max}) and eccentricity (e) of the orbits vs. pericentric distance (r_{peri}), while the right column shows the same as functions of apocentric distance (r_{apo}). Targets with $r_{apo} < 15$ kpc and $z_{max} < 3$ kpc are identified with "star" symbols (planar-like), targets with $r_{apo} < 30$ kpc are circle symbols (inner halo), and targets with $r_{apo} > 30$ kpc are shown as squares (outer halo). Distances are fixed at 4, 6, 8, and 10kpc (darkest to lightest points for each star respectively.)

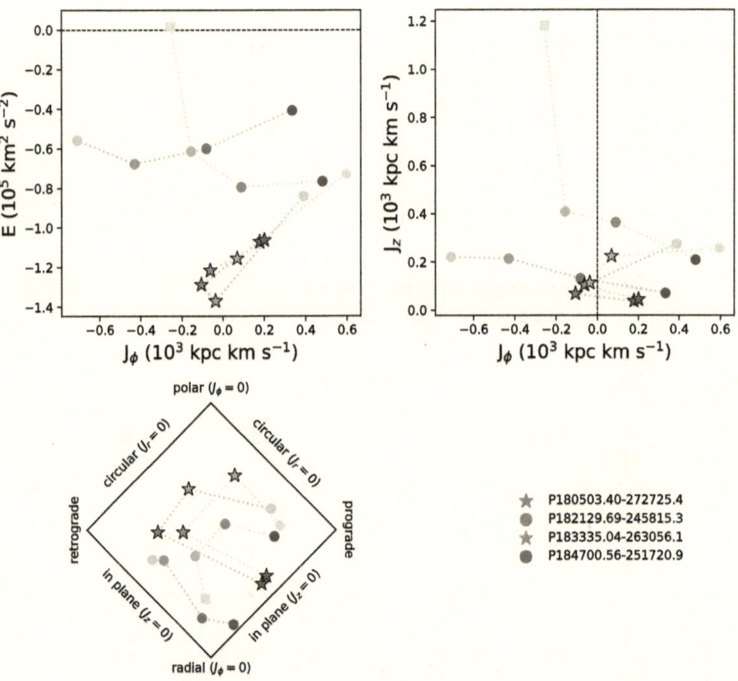

Figure 5.3 Orbital energies and rotational action vectors for the GRACES bulge sample. The orbital energy E is compared to the rotational action J_ϕ ($=L_z$) in the top left panel, and J_z is compared to J_ϕ in the top right. The bottom panel shows the action vector phase space in an alternative manner to highlight stars in polar vs. radial, retrograde vs. prograde, and circular or in-plane orbits. Prograde and retrograde regions are identified as having positive or negative J_ϕ. Symbols and colors are as in Fig. 5.2

a highly radial (large J_z) and unbound (E>0)orbit. Overall, I conclude that the uncertainties in distance are sufficient that we cannot rule out in situ formation for two stars.

The four stars presented here are a tiny sample of the $\sim$ 1300 VMP bulge candidates identified by Arentsen et al. (2020a). Collecting high resolution spectra for the remainder of these potentially remarkable stars cannot be done on a star-by-star basis. Future multi-object spectroscopic surveys like 4MOST (De Jong et al., 2019) and WEAVE (Dalton et al., 2014) will provide large samples of stars that will illuminate the inner regions of the Milky Way. These new beacons will allow us to explore the chemo-dynamic nature of the bulge and will provide much desired insight on the earliest generations of stars.

5.2 Conclusions

This book is a compendium of works centred around the spectroscopic study of Milky Way stellar relics. Follow up high-resolution CFHT ESPaDOnS and Gemini GRACES spectroscopy of 145 metal-poor candidates selected from the *Pristine* survey has uncovered 56 new very metal-poor stars with [Fe/H]< -2.5, 15 of which have [Fe/H]< -3.0. Compared to the expected metallicities derived from *Pristine* and SDSS photometry, these numbers translate to discovery success rates of 56% (56/100) and 33% (15/45), for the aforementioned metallicity bins. These rates are in excellent agreement with previous *Pristine* studies and are higher than what has been observed in other metal-poor surveys (Aguado et al., 2019b); an encouraging checkpoint as the *Pristine* survey continues forward in its' exploration of the metal-poor Galaxy. Furthermore, only Fe upper limits could be placed for three of the [Fe/H]< -3.0 EMP stars in the GRACES sample, warranting future follow-up observation.

These three stars could belong to the population of the oldest and most metal-poor fossils of the Milky Way, shedding crucial light on the earliest stages of star formation and Galactic evolution.

A model atmospheres analysis has been performed for these new very metal-poor stars, providing stellar parameters and chemical abundances for up to 19 elements (Li, Na, Mg, K, Ca, Sc, Ti, Cr, Mn, Fe, Ni, Cu, Zn, Y, Zr, Ba, La, Nd, Eu). The detailed chemical abundance analyses for the EMP stars increases the total number of known EMP stars with chemical abundance profiles by $\sim$ 5%. Most of the stars in the CFHT and Gemini samples show chemical abundance profiles that are similar to "normal" metal-poor halo stars, though a handful of stars show unique chemical signatures related to the CEMP stars, r-process enriched stars, neutron capture poor stars, stars with dwarf galaxy origins, and stars with the imprint of novel supernovae yields. The chemical diversity seen in this relatively small sample is exciting in the context of *Pristine*'s future discoveries.

The diversity of these stars are not only reflected in their chemistry, but also in their dynamics. The kinematics and orbits for these VMP and EMP stars were been calculated using radial velocities derived the HRS spectra an the *Gaia* DR2 proper motions and parallaxes. As the chemistry would suggest, a majority of the stars were found to be confined to the Galactic halo. Conversely, EMP stars with planar-like orbits, challenging our ideas of the formation of the Galactic plane. Additional stars with orbits which plunge deep into the Galactic centre have also been uncovered, and three stars are found to be chemo-dynamically linked to the *Gaia-Sequoia* accretion event. These exciting objects may provide key insight on the early chemo-dynamical formation of the Galaxy. This work is ongoing, including the high-resolution follow-up of metal-poor stars associated with the Galactic bulge, ensuring a productive career the next generation of *Pristine* graduate students.

I believe my responsibilities as an astronomer are not limited to the acquisition of new knowledge, but also to the development of techniques that may help facilitate the future growth of the field[1] The other projects contained in this book present two new data products that may assist current and future astronomers in the fields of stellar spectroscopy.

Considering the observational expense of high-resolution spectroscopic studies, medium-resolution studies are more suited for generating larger statistical samples over a shorter period of observing time. With that said, large data sets carry with them a need for fast and efficient data analysis tools. In Chapter 4, section 4.1, I presented a modified StarNet, a convolutional neural network designed to determine the stellar parameters T_{eff}, $\log g$, and [Fe/H] from medium-resolution stellar spectra. The spectra in the training and test set were augmented to include a range of S/N ratios and continuum normalization errors; data effects expected in medium-resolution IR spectra. While this medium-resolution StarNet model is capable of determining stellar parameters from medium-resolution spectra with comparable accuracy to its higher-resolution precursor, the need for the data augmentation steps in the training set should guide future spectroscopic surveys to minimize the errors in their spectra extraction and rectification routines. Finally, I presented an overview of emphNifty4Gemini, a python package which includes a data processing pipeline for Gemini NIFS observations. Compared to StarNet, NIFS is still very much a "by hand" data reduction tool, though its' capabilities to quickly produce flux and wavelength calibrated

---

[1]Though it should go without saying that our *most* important responsibility, as astronomers, is to interpret the knowledge that we collect about the Universe and then share it, in an open, inviting, and accessible manner, with the general public. I believe that increased scientific literacy among the human population is one of the most powerful and obtainable tools that we can use to combat the issues that we face as a species.

science-ready data cubes are remarkably valuable for NIFS users.

Future multi-object, high-resolution spectroscopic surveys like WEAVE(Dalton et al., 2012), 4MOST (De Jong et al., 2019), and SDSS-V(Kollmeier et al., 2017) (plus any future surveys on the next generation of telescopes), will inevitably outperform a classical, star-by-star, investigation of the Galaxy by many orders of magnitude. Coupled with improved data analysis techniques and the wide scale implementation of machine learning algorithms that will (hopefully) lighten the work load for astronomers, the future of Galactic chemical cartography indeed looks very bright. Though this book marks the end point of my formal engagement to astronomy, I am pleased to know there will always be exciting literature to read and I very much look forward to the revelations that *you all* uncover about our cosmic origins.

Appendix A

Examination of possible unbound stars in the CFHT ESPaDOnS Sample

Five stars were identified for a more careful dynamical analysis in Section 2.6.1. When their distances were determined from the Bayesian inference method, these stars had highly energetic and unbound orbits ($E/E_\odot < -0.5$), with $R_{apo} > 500$ kpc. These five stars are shown in Fig. A.1.

As two of these stars (Pristine_200.5298+08.9768 and Pristine_187.9785+08.7294) were found to have higher metallicities than had been adopted for the MIST stellar isochrone in the Bayesian inference method, their Bayesian inferred distances are assumed to be unreliable. Adopting their distances as 1/parallax from the Gaia DR2 database, sensible orbits and dynamical parameters are found for both stars. Furthermore, the parallax errors were small for both stars.

The orbit solution for a third star (Pristine_213.7879+08.4232) was also significantly improved by rejecting the Bayesian inferred distance in favour of the 1/parallax value. Again, the parallax error is small, and the resulting orbital properties are less peculiar. It is not clear why the Bayesian method did not work for this star, however we note that this was a target that was observed very early on and it is no longer in the *Pristine* survey catalogue. Investigating this star a bit further, we notice that the 1/parallax distance is closer than the Bayesian inferred distance, suggesting that the surface gravity for this star may be slightly higher ($\log g = 2.3$, may be closer to $\log g \sim 3$). In that case, we find small corrections to the abundances, such that $\Delta \log(\text{Fe II}) \sim +0.3$, bringing Fe II into much better agreement with Fe I. The impact on [Fe/H] for this star is negligible though since the iron abundance is dominated by the more numerous spectral lines of Fe I. Minor adjustments to the other

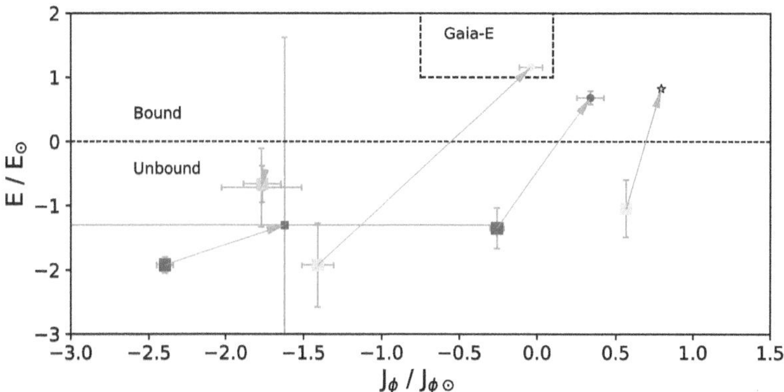

Figure A.1 Comparison of the orbital parameters for five stars with unbound orbits from the dynamical analysis in Section 2.6.1. Large symbols are the parameters using the distances from the Bayesian inference method, and small symbols are those from adopting 1/parallax from the Gaia DR2 database. Points coloured by [Fe/H]$_{Q6}$ using the same scheme same as in Fig. 2.16.

elements would have no signficant effect on the chemical analysis and interpretation of this star.

Finally, when examining the impact of the distances for the two stars Pristine_182.5364 ([Fe/H]=−1.6) and Pristine_181.4395 ([Fe/H]=−2.8), we find they always result in highly retrograde and unbound orbits. The orbit for the more metal-poor of these two stars is highly uncertain when determined from its parallax ($\Delta\pi/\pi = 0.45$). Interestingly though, this star is also one of the [Ba/Fe]-poor stars discussed in Section 2.5.4 as possibly accreted from an ultra faint dwarf galaxy. As a sanity check, we also calculated the orbits for all of the other metal-poor candidates in this analysis, but found only small offsets in their orbit and action parameters.

Appendix B

Spectra, upper limits, and line lists for the GRACES Sample

B.1   Spectra and Line Data

B.1.1   The 1D Spectra

The reduced 1D spectra for a subset of the stars in this paper (in T_{eff} and [Fe/H]) are shown in Figure B.1. Only the wavelength region used in this paper for the chemical analysis is shown; a zoomed in region from 470 to 680 nm. Each star and its metallicity are labelled, and the objects are sorted by effective temperature. The Balmer lines ($H\alpha$ and $H\beta$ are clear, as well as the Mgb lines. The atmospheric bands near 5850 and 6300 are clear in most stars, as well as some sky emission lines from an imperfect sky subtraction.These stars were not telluric cleaned due to inconsistent observations of telluric standards.

B.1.2   The world of (mostly) upper limits.

In higher metallicity stars, several spectral lines exist in the GRACES wavelength region of additional elements. We examine those regions, and provide a few abundances, but mostly upper limits only. The upper limits in this section do not provide useful constrains for nucleosynthetic interpretations, thus they are collected here in an Appendix, for completeness. These elements include the odd-elements (K, Sc, Mn, Cu), also Zn, and the neutron capture elements (Y, Zr, La, Nd). See Table D.23 for our line list and atomic data for these lines.

Sc and K abundances were measured in 1 and 5 stars, respectively. It is unclear if the slight increase in K with decreasing metallicity is astrophysically significant as the

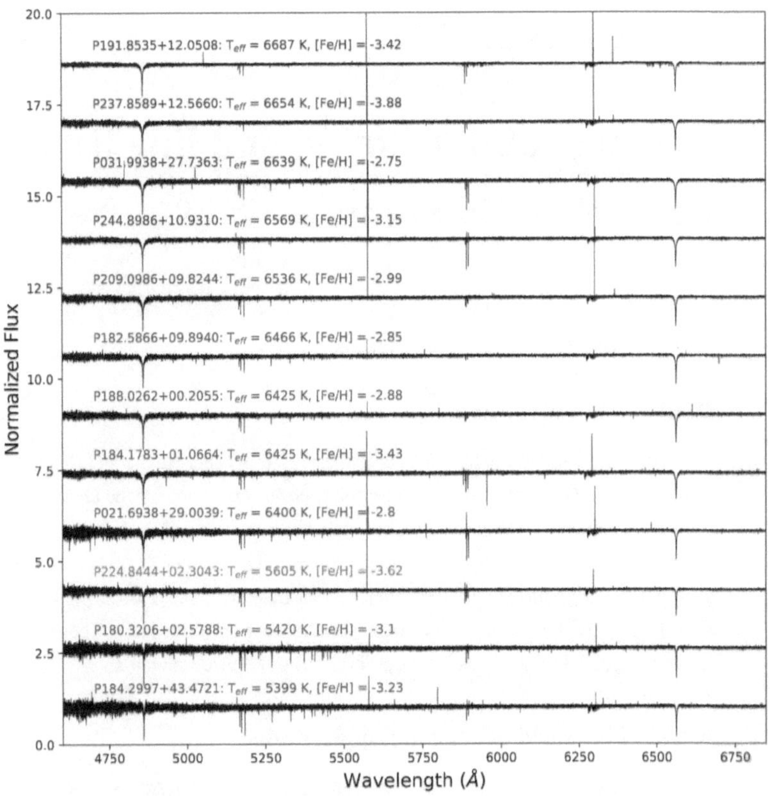

Figure B.1 Zoom in of the 1D spectra for a subset of the spectra analysed in this work, sorted by temperature. This region was selected for analysis as it has the highest SNR, is mostly free of telluric and sky lines, and contains spectral lines for many elements of interest.

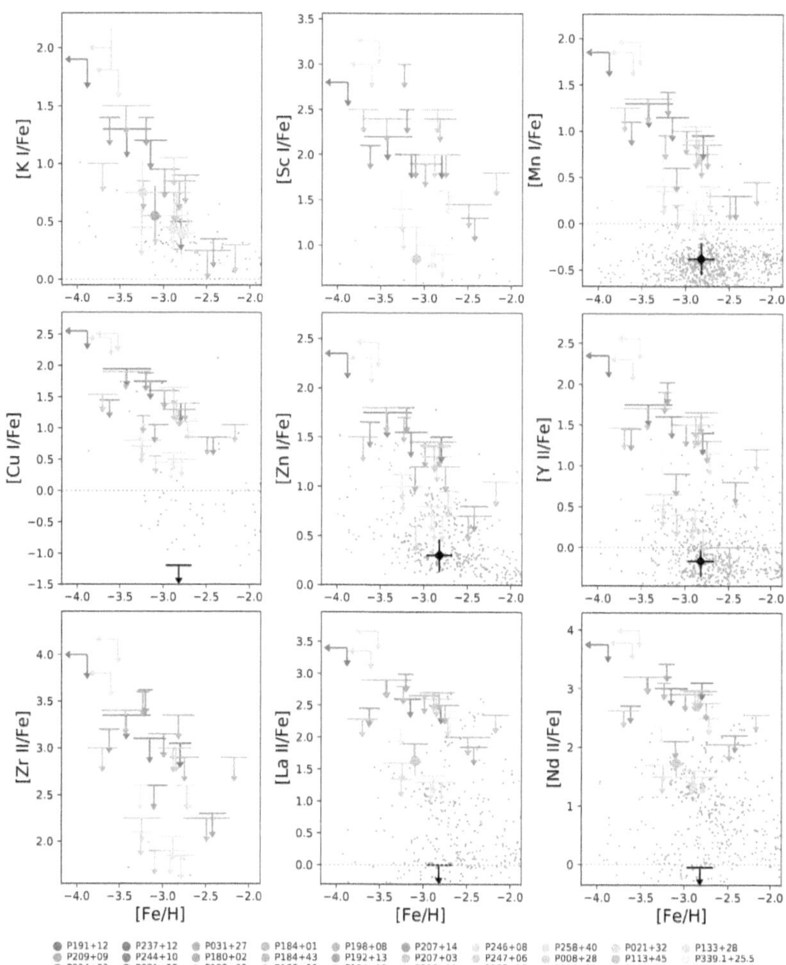

Figure B.2 Abundance plots for elements with upper limits only.

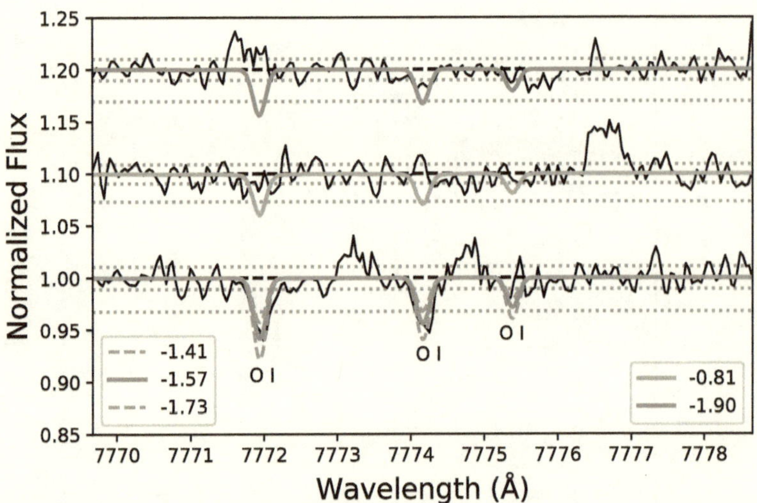

Figure B.3 Synthesized O lines for P207+14 (brown, bottom), P184+43 (blue, middle), and P192+13 (red, top). Labels the same as in Fig. 3.11.

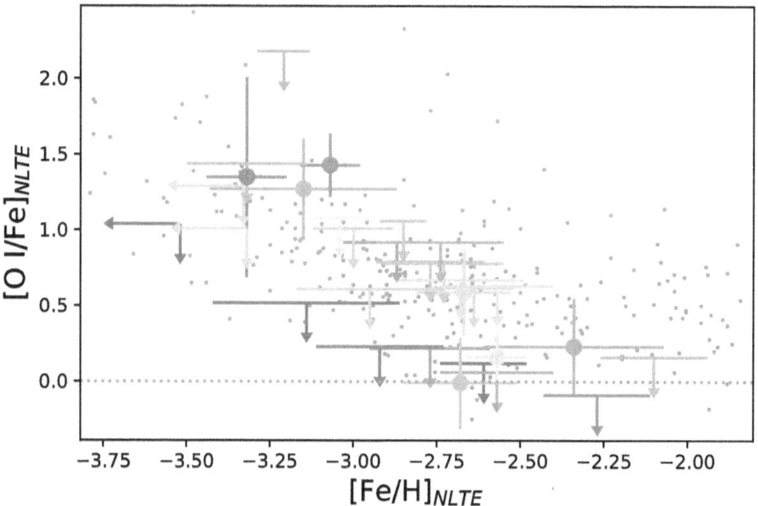

Figure B.4 NLTE [O I/Fe] vs. [Fe/H].

uncertainties for each star are large, but certainly some stars appear to be rich in K. The upper limits in Zn, Y, Nd, and La are in good agreement with the distribution of [X/Fe] in metal-poor stars in the Galaxy. One r-process rich star (P016+28; CEMP-s, discussed in the main text) also has enriched, and therefore measureable, La and Nd abundances.

Finally, O is included here since we analyse the strong $\lambda 7700$ triplet (see Fig. B.3; however, these lines are known to form over many layers in a stellar atmosphere, being very sensitive to small uncertainties in stellar parameters, and especially NLTE corrections. They are also in a region of significant telluric contamination. The O abundances for 7 stars where we could measure the OI triplet are shown in Fig. B.4, but should be taken with caution since they have very large uncertainties and no NLTE corrections.

B.1.3   Line List

A sample line list is provided in Table D.23, including the element, wavelength, excitation potential (χ, eV), and oscillator strengths ($\log gf$). The majority of this analysis was carried out using spectrum syntheses, thus the line abundance from each spectral feature is listed (rather than an equivalent width). These line abundances have been averaged together for the final abundance, per element, per star.

B.2   Orbital Analysis

Appendix for the orbital parameters calculated using the Bayesian inferred distances (rather than 1/parallax as in the main text.

Planar stars: P184+43 and P339.1+25.5 remain as planar-like candidates. P191+12 no longer a planar-like orbit, simply due to a slightly higher $|Z|_{max}$ now. New star P194+12 now appears to be in a planar-like orbit.

Sequoia stars: The three highly retrograde, high orbit energy stars from the 1/parallax analysis, P021+32, P031+27, and P133+28, do change positions in the Bayesian inferred distance analysis. Only P031+27 remains in the Gaia-Sequoia parameter space under the Bayesian inferred analysis, though it's low eccentricity may cause issues with this conclusion. P182+09, may be a new Gaia-Sequoia candidate based on the Bayesian inferred analysis - but it is also VERY CLOSE in the 1/parallax distances, so maybe we can just add it to BOTH orbital analyses. The other two are now unbound stars. This is very unlikely, and more of an indication of the problems encountered in the Bayesian analysis, likely due to the other uncertainties for EMP stars in the MIST/MESA isochrones.

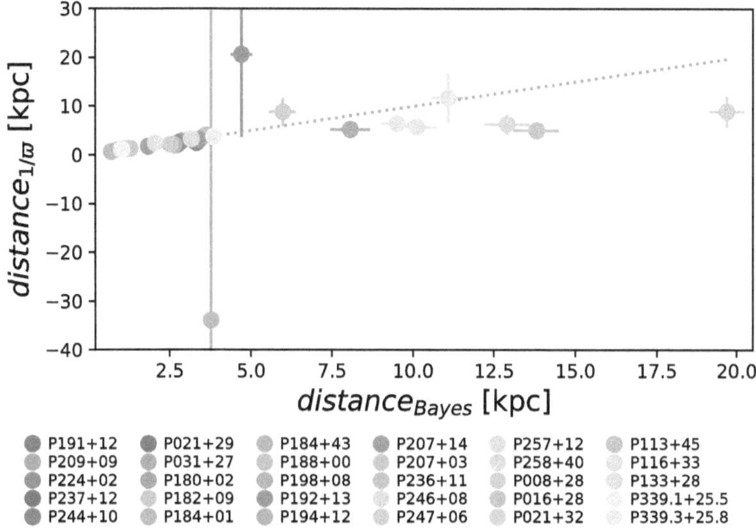

Figure B.5 Comparison of distances derived from inverting the parallax vs the distances derived from the Bayesian inference method. Though P198+08 has a negative distance from inverting the parallax, the orbital parameters for P198+08 in the main body of Chapter 3 were computed using a transformed gaussian PDF which was truncated to only positive distances.

Unbound Stars: Four stars in total appear to be unbound, based on the Bayesian analysis. Again, we do not expect to find this many unbound stars in a sample this small (30 stars). This contributes to our assumption that the 1/parallax distances are more reliable, and therefore used in the main text of this paper. Unfortunately, the choice of 1/parallax is inconsistent with the Bayesian inferred method for determining the stellar parameters. We discussed the uncertainties we predict in the stellar parameters due to likely uncertainties in the MIST/MESA isochrones in the main text of this paper.

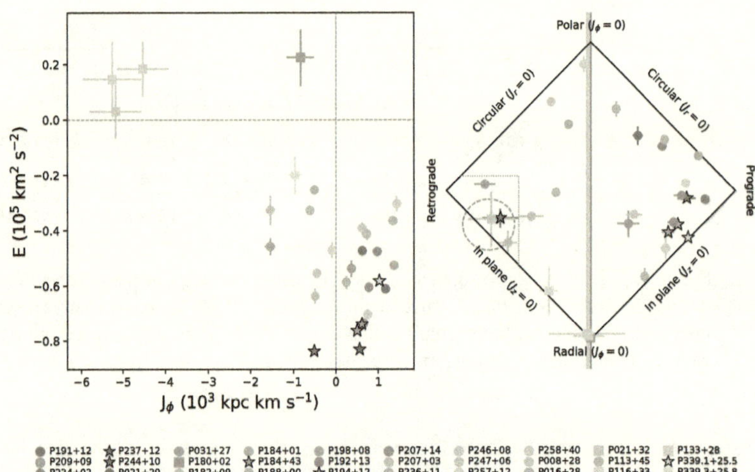

Figure B.6 Action vectors and energies for stars from 2018A-2019B datasets, but with distances inferred from the Bayesian analysis. Same symbols and labels as in Fig. 3.14.

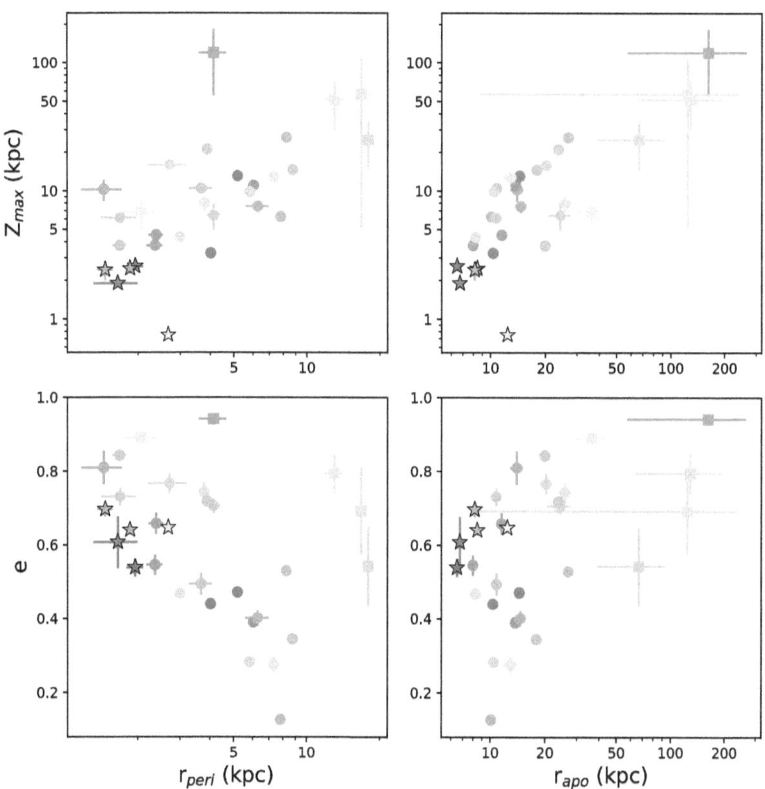

Figure B.7 Orbital elements for stars from 2018A-2019B datasets, but with distances inferred from the Bayesian analysis. Same symbols as comments as in Fig. B.6.

Appendix C

Gemini IRAF Tasks for *Nifty4Gemini*

C.1   nifcube

Overview

Create data cubes from NIFS 2D format files. NIFCUBE takes input from data output by either NSFITCOORDS or NSTRANSFORM and converts the 2D data images into data cubes that have coordinates of x, y, lambda.

Description

NIFCUBE is used to resample and combine spectra from the NIFS image slicer into three dimensional data cubes. It is a script wrapper around the more general task GEMCUBE. This wrapper handles details of the NIFS MEF format and various checks on the processing history. Additional details about the process of creating the data cubes may be found in the help for GEMCUBE.

The input NIFS spectra must be separate spectral images from each slice of the NIFS image slicer, stored as science extensions in a multiextension FITS (MEF) file, which have been mapped to world coordinates with the NSFITCOORDS task or resampled into rectified and calibrated slice data by the NSTRANSFORM task. The input files are checked for previous processing by NSFITCOORDS and a warning is issued if this is not the case. It is an error if no suitable input data is found and a warning is issued if the output already exists.

The input MEF files are generally organized as one file per exposure with each extension corresponding to one NIFS image slice on the sky. There may be science, error, and data quality extension versions. This task currently only utilizes and maps the science extensions. Each pixel in each science extension is mapped to one or more pixels in the output cube. These mapped pixels are combined to produce the output data cube or cubes. This process allows combining many input exposures, usually with some dithering or offsets, into a single data cube or creating a single data cube for each exposure. Whether multiple input files are combined into one data cube or separate data cubes are created for each input is determined by the number of input and output files specified. A shortcut for naming the output data cubes applies a prefix to each input filename. This shortcut is equivalent to specifying one output file name for each input filename.

The mapping from an input pixel to an output pixel depends on the input and output having world coordinates systems (WCS). The input WCS is set by the NSFITCO-ORDS task. Many of the details of this WCS are described in help for that task. The NSTRANSFORM task may resample the original sampling of the slice images into rectified slice images with a new, simpler, WCS. The result of NSTRANSFORM may also be used to create data cubes though generally the results of NSFITCOORDS would be used directly to eliminate an intermediate interpolation step.

The output WCS is set either using a reference data cube or using NIFCUBE parameters and certain default behaviors. In the former case the output WCS is specified by a reference image name. The image must have a three-dimensional WCS. Generally this might be a data cube, possibly produced previously by NIFCUBE, that one wants to match. However, it could be a template WCS header in an image without pixels.

Note that the reference WCS does not specify the output coverage but only which axes correspond to wavelength and space, their orientation, and the pixel scales. The coverage is specified by the range of data in the input and by the wmin and wmax task parameters.

If a WCS reference is not supplied then the output data cube will have a sampling specified by task parameters. The orientation of the axes is set by the waxis, saxis, and taxis parameters. The default is to place wavelength along the third or "z" axis, the spatial sampling of a slice along the second or "y" axis, and the sampling across the slices along the first or "x" axis. The output pixel spatial scales are constrained to be the same; i.e. the spatial footprint of the pixels is square even though the instrumental sampling is not. The spatial size of the pixels may be explicitly set by the sscale parameter with a value of "INDEF" selecting the highest resolution in the input data. The wavelength sampling is similarly set by the wscale parameter with a default to the the highest spectral resolution in the input data. There is a special syntax to select the default scales but force the world coordinate increase with pixel coordinate, "+INDEF", or decrease pixel coordinate, "-INDEF". As with the WCS reference method, the coverage is set by the range of the input data except that a range of wavelength may be specified if desired.

Because the output coverage is set to include all the input data, with the possible exception of limiting the wavelength range of interest, there will be output pixels along the edges which do not correspond to any input pixels. The values of these output pixels is set by the blank parameter. One is free to use this value either to produce minimal impact for later applications or to provide a special value outside the range of the input data for later identification as out of bounds regions.

The logfile and verbose parameters provide for displaying and/or recording general information documenting the effects of this task.

The drizscale parameter is described further below.

3D DRIZZLE This section summarizes the algorithm producing the output pixel values from the input pixel values. The algorithm is an extension of the algorithm first developed for 2D resampling (Fruchter & Hook (2002)) which goes by the common name "drizzle".

Each input pixel maps to a rectangular volume (voxel) with two spatial dimensions, a dispersion dimension, and some orientation. In our case the wavelength axis is assumed to be orthogonal to the spatial axes. Though the input data is two dimensional the third dimension is implicit in the slice which has a pixel width defined by the the effective slit size.

In the drizzle algorithm the value of the input voxel is apportioned to the output voxels that it overlaps by the volume of the overlap. There may be multiple input voxels that overlap the same output voxel either from the same exposure or from multiple exposures when combining data. In either case the apportioned value from the input voxels are accumulated along with the volume fraction. The accumulated fractions are used to normalize the final output data cube value.

Note that this method preserves the input pixel values independent of the output pixel sampling. This means that it is effectively preserving the input "surface brightness" and not the total flux. However, by the nature of this method the sum of all the pixel values which map to the output data cube is rigorously conserved.

This method is a type of linear interpolation which, like all resampling methods, degrades the resolution to some extent. However, by treating the input pixels as if they were smaller this effect can be minimized. The drizscale parameter allows the input voxel volume to be shrunk (or expanded) by specified factors. This shrinking can be uniform by specifying a single scale value or can be applied separately for each world coordinate dimension.

Because shrinking the input pixels can leave gaps in the output cube this only makes sense when combining multiple exposures that have been offset by a non-integer number of input pixels. It is also generally used when the output is created with finer sampling than the input. The Fruchter and Hook reference provides some guidelines and discussion about how drizzling might be used.

Note that this capability will only be useful when the input data supports WCS that incorporate information about offsets, relative orientations, and dithering used during the observations. This is not yet developed for NIFS though it will be added as soon as possible.

Important Parameters

reference = ""

> A reference cube whose WCS is matched in the output cubes. This does not mean the size of the cubes will be the same, but where the output and reference cubes overlap in world coordinates they will have the same pixel sampling to allow simple combining. Note that one may use a WCS image which has only a header and no data.

waxis, saxis, taxis = 3,2,1

> The output cube axes for wavelength, the spatial component along the slices, and the spatial component across the slices respectively.

wmin, wmax = INDEF

> The minimum and maximum wavelengths, in Angstroms, for the output cubes. If INDEF is specified the range of the input spectra provide the limits.

If no WCS reference is specified the following parameters are used to define the output cube sampling.

wscale, sscale = INDEF, "0.05"

The output pixel sampling for the wavelength axis, in Angstroms per pixel, and the spatial axes, in arc seconds per pixels, respectively. Note that the two spatial axes will have the same output pixel size, that is square pixels on the sky, even though the input slice sampling is different along and across the slices. If INDEF is specified the highest resolution sampling in the input data is used. If +INDEF or -INDEF is specified then the world coordinate will increase or decrease with increasing pixel coordinate in the output cube.

drizscale = "1."

The volume of the input pixels in world coordinates is scaled by this factor. The value may be a single value that applies to all threes axes or three values for the two spatial axes and the wavelength axis respectively.

blank = 0.

Output pixel value when no input data contribute to that pixel.

Appendix D

Data Tables from Chapters 2 and 3

The data tables from Chapters 2 and 3 are collected here, to ease the presentation of the scientific results in the aforementioned chapters.

Table D.1: Metal-Poor Targets (115) from the original *Pristine* survey footprint. 59 stars were selected that have a *Pristine* metallicity $[Fe/H]_P < -2.5$ and SDSS *(g-i)* and *(g-r)* calibrations with probabilities for both $[Fe/H]_{Pgi}$ and $[Fe/H]_{Pgr} < -2.25$ greater than 80% $(dFe_P = 1\sigma)$. For 10 stars, only the SDSS *(g-r)* calibration was available, as noted (where we flagged SDSS *i* saturation).. Targets no longer in the *Pristine* catalogue, or with $[Fe/H]_P > -2.5$, are noted. The CaHK and SDSS *ugri* magnitudes are dereddened using the E(B-V) values from the (Schlegel et al., 1998) maps. The V and I magnitudes are not dereddened (i.e., observer units). The SDSS colour temperature (T_{SDSS}) averages the dwarf and giant solutions (where $dT = 1\sigma$). The CFHT semester program labels are in the comments.

RA_{SDSS} (deg)	DEC_{SDSS} (deg)	V	I	E(B-V)	$CaHK_0$	u_0	g_0	r_0	i_0	T_{SDSS} (K)	dT (K)	$[Fe/H]_P$	dFe_P	COMM
High probability for $[Fe/H]_P \leq -2.5$ in the *Pristine* catalogue:														
180.2206	09.5683	14.92	13.99	0.021	15.822	16.439	15.172	14.625	14.392	5202.6	17.6	-2.82	0.02	16BC,17AC
181.2243	07.4160	14.95	14.33	0.015	15.361	15.949	15.044	14.804	14.707	6261.5	5.8	-2.78	0.01	17AC002
181.3464	11.6698	14.18	13.42	0.033	14.782	15.293	14.392	13.841	13.757	5504.8	15.6	-3.82	0.09	16AC031
181.4395	01.6294	14.67	13.66	0.020	15.562	16.345	14.961	14.357	14.078	5011.6	18.5	-3.82	0.09	16BC008
181.6953	13.8075	14.52	13.73	0.031	15.135	15.701	14.692	14.228	14.087	5572.5	15.0	-3.37	0.06	18BC018
182.5364	00.9431	14.97	14.33	0.029	15.395	15.975	15.036	14.771	14.678	6201.8	6.8	-2.46	0.05	17AC002
183.6849	04.8619	14.96	14.37	0.018	15.285	15.823	15.025	14.822	14.735	6398.9	3.2	-3.13	0.02	18BC018
185.4110	07.4777	14.72	14.07	0.022	15.168	15.657	14.826	14.523	14.430	6096.2	8.6	-2.94	0.04	17AC002
187.9785	08.7294	14.93	14.18	0.018	15.640	16.179	15.178	14.661	14.552	5525.0	15.5	-3.33	0.23	16BC008
189.9449	11.5534	14.39	13.74	0.035	14.729	15.321	14.416	14.188	14.081	6267.2	5.7	-2.81	0.02	18BC018
190.6313	08.5137	14.92	14.16	0.021	15.601	16.402	15.101	14.673	14.534	5660.9	14.2	-2.76	0.12	17AC002
193.1159	08.0557	14.56	13.89	0.026	15.030	15.489	14.646	14.361	14.253	6104.5	8.5	-2.46	0.01	17AC002
193.5533	11.5037	14.18	13.37	0.033	15.040	15.407	14.440	13.818	13.717	5316.7	17.0	-2.83	0.46	16AC031
193.8390	11.4150	14.75	13.61	0.030	15.962	16.671	15.055	14.364	14.017	4757.8	18.9	-2.91	0.12	18BC018
196.3755	08.5138	14.65	13.70	0.029	15.577	16.175	14.899	14.326	14.085	5137.4	17.9	-2.84	0.01	17AC002
196.4117	14.3176	14.57	13.67	0.028	15.534	15.935	14.870	14.193	14.038	5103.8	18.1	-3.04	0.36	16AC096
196.5453	12.1211	14.71	14.16	0.026	15.020	15.626	14.739	14.562	14.506	6572.4	0.5	-2.79	0.03	18BC018

Continued on next page

Table D.1 – *continued from previous page*

RA_{SDSS} (deg)	DEC_{SDSS} (deg)	V	I	E(B-V)	$CaHK_0$	u_0	g_0	r_0	i_0	T_{SDSS} (K)	dT (K)	$[Fe/H]_P$	dFe_P	COMM
198.5486	11.4123	14.27	13.65	0.023	14.619	15.223	14.341	14.093	14.011	6281.6	5.4	-3.15	0.01	16AC031
200.0999	13.7228	14.94	13.84	0.025	16.185	16.864	15.267	14.570	14.250	4790.0	19.0	-2.81	0.14	18BC018
200.5298	08.9768	15.15	14.32	0.027	15.821	16.575	15.342	14.870	14.695	5478.3	15.9	-3.30	0.01	16BC008
201.1158	15.4381	14.31	13.50	0.020	15.115	15.893	14.538	14.039	13.884	5463.0	16.0	-2.61	0.11	18BC018
201.3732	08.4513	14.65	13.92	0.018	15.192	16.047	14.834	14.432	14.305	5759.5	13.1	-3.34	0.05	16BC,18BC
201.8711	07.1810	14.76	14.15	0.030	15.045	15.592	14.799	14.562	14.495	6357.4	3.9	-3.33	0.01	18BC018
202.3435	13.2291	14.24	13.48	0.021	14.706	15.263	14.386	14.027	13.864	5769.4	13.1	-3.80	0.08	16BC,18BC
203.2831	13.6326	14.85	13.88	0.024	15.765	16.417	15.103	14.536	14.277	5115.0	18.1	-3.01	0.07	16BC008
204.9008	10.5513	14.88	14.31	0.028	15.119	15.745	14.886	14.717	14.657	6584.9	0.8	-3.23	0.09	17AC002
205.1342	13.8234	14.61	13.84	0.023	15.313	15.955	14.788	14.368	14.212	5639.6	14.4	-2.50	0.06	17AC002
205.8131	15.3832	14.64	14.08	0.032	14.863	15.482	14.659	14.458	14.408	6516.8	0.7	-3.48	0.05	18BC018
206.3487	09.3099	14.29	13.71	0.027	14.657	15.204	14.347	14.110	14.055	6393.0	3.3	-2.64	0.09	18BC018
208.0798	04.4266	14.54	13.79	0.028	14.988	23.848	14.664	14.308	14.158	5809.0	12.6	-3.72	0.06	18BC018
209.9364	15.9251	14.85	14.28	0.020	15.175	15.781	14.894	14.706	14.637	6498.4	1.1	-2.85	0.02	18BC018
210.0166	14.6289	14.57	13.70	0.018	15.390	15.908	14.779	14.320	14.100	5408.9	16.4	-2.53	0.07	16AC031
210.8632	08.1797	14.50	13.72	0.023	15.161	15.653	14.667	14.260	14.098	5656.2	14.3	-2.70	0.02	18BC018
213.2814	14.8983	14.55	13.94	0.018	14.935	15.819	14.637	14.394	14.308	6284.5	5.3	-2.93	0.01	17AC002
214.5556	07.4670	14.64	14.04	0.030	15.007	15.584	14.695	14.442	14.382	6331.0	4.5	-2.70	0.10	18BC018
217.3861	15.1650	14.61	14.03	0.025	14.937	15.546	14.660	14.436	14.378	6422.8	2.7	-2.94	0.05	18BC018
217.5786	14.0379	14.49	13.51	0.029	15.433	16.086	14.750	14.155	13.901	5072.5	18.2	-3.04	0.02	16BC008
217.6443	15.9633	14.85	14.26	0.031	15.171	15.756	14.879	14.666	14.597	6422.8	2.7	-2.79	0.05	18BC018
218.4622	10.3683	14.90	14.26	0.023	15.280	16.105	14.986	14.716	14.622	6184.9	7.2	-3.22	0.03	17AC002
228.4607	08.3553	14.85	14.28	0.030	15.179	15.709	14.876	14.672	14.620	6501.5	1.1	-2.64	0.08	17AC002
228.6557	09.0914	14.79	14.24	0.031	15.086	15.664	14.818	14.613	14.569	6523.0	0.6	-2.96	0.04	17AC002
228.8159	00.2222	14.76	14.13	0.053	15.023	15.581	14.719	14.497	14.424	6384.0	3.5	-2.71	0.06	17AC002

Continued on next page

Table D.1 – *continued from previous page*

RA$_{SDSS}$	DEC$_{SDSS}$	V	I	E(B-V)	CaHK$_0$	u$_0$	g$_0$	r$_0$	i$_0$	T$_{SDSS}$	dT	[Fe/H]$_P$	dFe$_P$	COMM
(deg)	(deg)									(K)	(K)			
233.5730	06.4702	14.55	13.56	0.044	15.445	16.062	14.750	14.166	13.920	5107.5	18.1	-2.83	0.01	17AC002
235.1449	08.7463	14.58	13.87	0.042	14.980	15.533	14.627	14.327	14.200	6012.6	9.9	-3.02	0.07	16BC008
236.1068	10.5311	14.54	13.45	0.050	15.689	16.383	14.775	14.097	13.806	4866.0	18.8	-2.69	0.08	16AC031
237.8246	10.1426	15.08	14.21	0.052	15.656	16.223	15.169	14.740	14.543	5525.0	15.4	-3.12	0.07	18BC018
240.4216	09.6761	14.77	13.85	0.040	15.473	16.084	14.924	14.438	14.211	5337.4	16.9	-3.14	0.12	16AC,16BC
245.5747	06.8844	14.95	14.09	0.064	15.478	16.038	15.000	14.568	14.391	5563.4	15.1	-3.11	0.02	16BC,18BC
245.8356	13.8777	13.86	12.97	0.045	14.547	15.069	13.987	13.522	13.316	5426.0	16.2	-2.81	0.04	16AC031
246.5144	05.9826	14.47	13.51	0.068	15.110	16.021	14.591	14.020	13.815	5210.4	17.5	-3.73	0.06	18BC018
246.8588	12.3193	14.61	13.67	0.057	15.393	15.843	14.730	14.207	13.995	5292.2	17.2	-2.52	0.02	16AC096
248.4959	15.0776	14.57	13.58	0.058	15.365	16.025	14.698	14.168	13.910	5187.1	17.7	-2.76	0.09	16AC096
250.6963	08.3743	14.52	13.52	0.085	15.233	15.836	14.567	14.025	13.793	5214.4	17.6	-2.70	0.02	16AC096
251.4082	12.3657	14.85	13.80	0.055	15.727	16.394	15.034	14.423	14.142	4995.7	18.5	-3.33	0.12	17AC002
252.1639	15.0648	14.43	13.38	0.069	15.417	16.018	14.578	13.938	13.690	5002.7	18.5	-2.45	0.07	16AC096
253.8582	15.7240	14.88	13.87	0.090	15.579	16.157	14.915	14.367	14.134	5200.6	17.6	-2.74	0.02	18BC018
254.5469	10.9129	14.40	13.49	0.077	14.990	15.447	14.403	13.978	13.774	5518.3	15.5	-2.41	0.09	16AC031
255.2671	14.9711	14.41	13.72	0.081	14.611	15.170	14.292	14.048	13.966	6293.2	5.3	-2.68	0.05	16AC031
255.5555	10.8612	14.76	13.94	0.074	15.183	15.850	14.746	14.370	14.215	5747.4	13.3	-2.92	0.02	16AC096

Used the SDSS-gr solution only for [Fe/H]$_P \leq -2.5$ in the *Pristine* catalogue:

RA$_{SDSS}$	DEC$_{SDSS}$	V	I	E(B-V)	CaHK$_0$	u$_0$	g$_0$	r$_0$	i$_0$	T$_{SDSS}$	dT	[Fe/H]$_P$	dFe$_P$	COMM
180.3789	00.9470	14.15	13.43	0.019	14.900	15.332	14.341	13.910	13.798	5718.2	13.6	-2.48	...	17AC002
181.3699	11.7636	14.19	13.38	0.037	15.164	15.461	14.411	13.838	13.720	5383.4	16.6	-2.50	...	16AC031
188.1264	08.7740	14.22	13.42	0.018	15.162	15.593	14.515	13.917	13.798	5329.0	16.9	-3.28	...	17AC002
200.7620	09.4375	14.51	13.75	0.025	15.222	16.010	14.677	14.255	14.119	5682.2	13.9	-2.49	...	17AC002
229.1219	00.9089	14.76	14.15	0.047	15.040	15.607	14.725	14.525	14.461	6477.1	1.6	-2.72	...	17AC002
235.7578	09.0000	14.92	14.34	0.041	15.257	15.819	14.916	14.704	14.651	6474.1	1.7	-2.56	...	17AC002

Continued on next page

Table D.1 – *continued from previous page*

RA_{SDSS} (deg)	DEC_{SDSS} (deg)	V	I	E(B-V)	$CaHK_0$	u_0	g_0	r_0	i_0	T_{SDSS} (K)	dT (K)	$[Fe/H]_P$	dFe_P	COMM
237.9600	15.4022	14.55	13.96	0.048	14.827	15.329	14.510	14.314	14.264	6532.2	0.4	-2.69	...	16AC031
245.4387	08.9954	14.77	14.12	0.056	15.109	15.613	14.747	14.491	14.415	6275.9	5.5	-2.47	...	17AC002
248.4394	07.9229	14.41	13.74	0.074	14.678	15.182	14.320	14.066	13.999	6307.7	5.0	-2.49	...	16AC031
254.5207	15.4969	13.93	13.16	0.086	14.718	15.096	14.030	13.396	13.378	5467.4	15.9	-3.23	...	16AC031

Low probability for $[Fe/H]_P \leq -2.5$ in the current *Pristine* catalogue:

RA_{SDSS} (deg)	DEC_{SDSS} (deg)	V	I	E(B-V)	$CaHK_0$	u_0	g_0	r_0	i_0	T_{SDSS} (K)	dT (K)	$[Fe/H]_P$	dFe_P	COMM
182.8512	14.1595	14.19	13.14	0.042	15.258	15.959	14.396	13.823	13.521	5025.7	18.4	-2.3	...	16AC031
196.6015	15.6768	14.44	13.67	0.029	15.355	15.780	14.602	14.167	14.028	5644.4	14.4	-1.2	...	16AC096
197.9861	12.3577	14.98	14.30	0.029	15.586	16.030	15.056	14.765	14.657	6088.0	8.7	-1.4	...	18BC018
209.2123	01.5275	14.45	13.72	0.034	15.013	15.504	14.546	14.202	14.071	5887.2	11.6	-2.3	...	17AC002
209.7175	10.8612	14.55	13.92	0.025	14.996	15.460	14.627	14.371	14.277	6224.4	6.5	-2.4	...	16AC031
210.7504	12.7744	14.50	13.34	0.027	16.054	16.702	14.869	14.089	13.749	4639.0	19.0	-2.0	...	16AC096
211.2766	10.3280	14.71	14.05	0.019	15.223	15.807	14.823	14.525	14.424	6088.0	8.7	-2.4	...	17AC002
212.5834	10.5365	14.42	13.74	0.023	14.970	15.395	14.540	14.217	14.108	5999.3	10.1	-2.4	...	17AC002
215.6121	15.0163	14.20	13.32	0.024	15.082	15.560	14.403	13.926	13.713	5385.5	16.6	-2.2	...	16AC031
216.1245	10.2135	14.49	13.65	0.027	15.257	15.773	14.666	14.219	14.029	5500.5	15.7	-2.4	...	17AC002
223.5273	11.1353	14.28	13.18	0.032	15.681	16.306	14.596	13.884	13.577	4786.8	18.9	-2.1	...	16AC096
227.2885	01.3598	14.45	13.70	0.052	15.328	15.720	14.513	14.127	14.011	5818.9	12.4	-0.5	...	16BC008
229.0409	10.3020	14.45	13.39	0.039	15.709	16.330	14.701	14.054	13.768	4925.4	18.8	-1.9	...	16AC031
232.6956	08.3392	14.97	14.18	0.044	15.561	16.051	15.052	14.681	14.513	5727.9	13.5	-2.4	...	17AC002
233.9312	09.5596	14.84	14.06	0.037	15.466	15.931	14.949	14.564	14.410	5727.9	13.5	-2.3	...	17AC002
234.4398	12.6286	14.46	13.61	0.039	15.604	15.979	14.635	14.132	13.961	5419.6	16.3	-0.8	...	16AC096
235.9710	09.1864	14.61	13.99	0.041	14.974	15.475	14.611	14.380	14.307	6357.4	3.9	-2.4	...	17AC002
236.0763	04.0496	14.33	13.54	0.069	15.264	15.653	14.361	13.936	13.820	5723.0	13.5	-0.4	...	16AC096
241.1186	09.4156	14.48	13.80	0.046	14.896	15.395	14.499	14.221	14.117	6134.8	8.1	-2.2	...	17AC002

Continued on next page

Table D.1 – *continued from previous page*

RA_{SDSS} (deg)	DEC_{SDSS} (deg)	V	I	E(B-V)	$CaHK_0$	u_0	g_0	r_0	i_0	T_{SDSS} (K)	dT (K)	$[Fe/H]_P$	dFe_P	COMM
No longer in the *Pristine* catalogue:														
180.7918	03.4085	13.18	12.12	0.022	...	15.512	13.753	12.652	12.501	4494.0	...	-2.8	...	16BC008
182.1679	03.4771	13.61	12.80	0.017	...	15.397	14.199	13.099	13.138	4735.0	...	-3.8	...	16BC008
182.5089	03.7165	13.27	12.74	0.016	...	14.590	13.610	12.945	13.067	5736.0	...	-4.0	...	16BC008
187.8508	13.4559	13.50	13.16	0.022	...	14.599	13.920	13.077	13.421	...	...	-2.5	...	16AC031
190.5804	12.8577	13.91	13.39	0.026	...	15.173	14.326	13.473	13.675	...	...	-2.5	...	16AC096
192.2112	15.9262	14.42	13.47	0.022	...	16.268	14.836	14.001	13.850	...	...	-2.3	...	16AC031
192.8531	15.8198	13.83	13.53	0.020	...	15.131	14.270	13.405	13.786	...	...	-2.3	...	16AC031
199.9279	08.3815	13.69	13.15	0.020	...	15.451	14.101	13.293	13.453	5514.0	...	-3.9	...	16BC008
207.9961	01.1795	13.84	13.25	0.026	...	15.156	14.302	13.373	13.545	5258.0	...	-3.9	...	16BC008
211.7174	15.5515	14.52	13.48	0.014	...	16.490	14.887	14.178	13.906	...	...	-2.5	...	16AC096
213.7879	08.4232	14.81	13.90	0.026	...	16.015	14.995	14.546	14.299	5396.0	13.4	-3.3	...	16BC008
215.6783	07.6929	14.09	13.61	0.025	...	14.536	13.890	13.638	15.348	5527.0	...	-3.5	...	16BC008
218.4256	07.5213	14.19	13.76	0.020	...	15.539	14.675	13.743	14.034	...	...	-2.6	...	16AC096
229.8912	00.1105	14.10	13.43	0.055	...	15.457	14.438	13.569	13.681	5296.0	...	-3.3	...	16BC008
230.4662	06.5251	14.17	13.50	0.034	...	15.392	14.534	13.732	13.803	5348.0	...	-3.5	...	16BC008
231.0319	06.4866	14.26	13.67	0.037	...	15.556	14.627	13.794	13.948	5468.0	...	-3.5	...	16BC008
232.8039	06.1178	14.76	14.04	0.042	...	16.594	15.027	14.352	14.351	5440.0	13.0	-2.6	...	16BC008
236.4846	10.6903	14.29	13.84	0.044	...	15.367	14.579	13.842	14.084	...	...	-3.3	...	16AC031
236.7139	09.6084	14.44	13.07	0.036	...	15.650	14.533	14.185	13.541	4851.0	...	-3.7	...	16BC008
237.8344	10.5901	14.63	13.70	0.045	...	15.840	14.743	14.309	14.059	5402.0	13.4	-2.5	...	16AC096
240.0340	13.8279	14.57	13.55	0.044	...	16.250	14.791	14.179	13.914	...	...	-2.5	...	16AC096
245.1095	08.8947	14.18	13.58	0.056	...	15.395	14.459	13.681	13.824	5567.0	...	-3.1	...	16BC008
250.8789	12.1101	14.14	12.69	0.046	...	15.662	14.273	13.808	13.154	...	...	-2.5	...	16AC031

Continued on next page

Table D.1 – *continued from previous page*

RA_{SDSS}	DEC_{SDSS}	V	I	E(B-V)	$CaHK_0$	u_0	g_0	r_0	i_0	T_{SDSS}	dT	$[Fe/H]_P$	dFe_P	COMM
(deg)	(deg)									(K)	(K)			
252.4199	12.6477	14.43	13.44	0.045	...	15.498	14.487	14.151	13.822	...	...	-2.4	...	16AC031
252.4908	15.2984	14.37	13.14	0.059	...	15.424	14.353	14.077	13.548	...	...	-2.6	...	16AC096
254.0653	14.2694	13.39	12.99	0.060	...	14.523	13.657	12.885	13.187	...	...	-2.5	...	16AC031
254.7759	13.8207	13.67	13.16	0.064	...	14.848	14.026	13.068	13.360	...	...	-3.2	...	16AC096

Table D.2: Gaia DR2 parallaxes, and the derived distances (D), temperatures (T), and surface gravities (log g) from the Bayesian inference method (see Section 2.4.1, assuming $[Fe/H]_P$). Corresponding uncertainties are listed as dpar, dD, dT, and dlg, respectively. Metallicities are from our "Quick Six" analysis (see Section 2.4.2) as the individual ion abundances, the weighted average $[Fe/H]_{Q6}$, the standard deviation σFe_{Q6}, and total number of lines used. Radial velocities are in the heliocentric reference frame. For four targets, a second distance (D2 in Com) satisifies the Bayesian inference analyais but does not significantly affect the stellar parameters. For four other targets, a dwarf or giant solution has equal probability, and we examine both solutions here independently. Stars no longer in the *Pristine* catalogue have been excluded since they are not metal-poor targets, with only one exception (RA=213.7879, DEC=+08.4232, noted as **).

RA_{SDSS}	DEC_{SDSS}	par	dpar	D	dD	T	dT	log(g)	dlg	RV	Fe I	Fe II	$[Fe/H]_{Q6}$	σFe_{Q6}	N,Com
		μas	μas	kpc	kpc	K	K	dex	dex	$km\,s^{-1}$	dex	dex	dex	dex	
High probability for $[Fe/H]_P \leq -2.5$:															
180.2206	+09.5683	0.09	0.05	12.08	0.47	5070.6	20.9	1.86	0.05	23.0	4.54	4.31	-3.04	0.18	6
181.2243	+07.4160	0.47	0.05	2.28	0.07	6454.9	99.9	3.81	0.06	-147.0	4.46	4.18	-3.16	0.16	5
181.3464	+11.6698	1.23	0.03	0.60	0.01	6208.3	17.1	4.57	0.01	12.0	7.55	6.86	-0.22	0.15	5
181.4395	+01.6294	0.08	0.05	16.76	0.29	4934.9	8.2	1.41	0.02	206.0	4.83	4.57	-2.76	0.17	6
181.6953	+13.8076	0.77	0.03	2.74	0.11	5708.0	69.0	3.24	0.04	79.0	6.23	6.31	-1.24	0.18	5
182.5364	+00.9431	0.45	0.06	2.21	0.12	6370.1	202.0	3.81	0.11	222.0	5.96	5.83	-1.59	0.1	5
183.6849	+04.8620	0.95	0.28	1.06	0.05	6491.0	42.0	4.44	0.03	40.0	4.38	4.3	-3.16	0.17	4, D2
185.4110	+07.4777	1.08	0.04	0.8	0.02	6304.0	26.0	4.53	0.01	176.0	5.93	5.48	-1.75	0.28	5
187.9785	+08.7294	0.74	0.04	5.84	0.59	5418.0	42.6	2.66	0.11	-55.0	6.68	6.58	-0.86	0.2	5
189.9449	+11.5535	0.61	0.03	1.69	0.03	6491.0	133.0	3.83	0.04	32.0	4.78	4.5	-2.82	0.12	6
190.6313	+08.5137	0.58	0.05	3.27	0.21	5591.6	35.4	3.21	0.07	-64.0	6.72	6.63	-0.82	0.25	5
193.1159	+08.0557	1.13	0.03	0.79	0.02	6359.3	22.6	4.49	0.01	43.0	5.6	5.6	-1.9	0.34	6
193.5533	+11.5037	1.42	0.03	0.56	0.01	6078.8	17.3	4.6	0.01	16.0	6.84	6.87	-0.65	0.15	5
193.8390	+11.4150	-0.02	0.04	19.12	0.4	4764.0	32.0	1.22	0.03	4.0	4.73	4.54	-2.83	0.12	6
196.3755	+08.5138	0.07	0.03	11.86	0.44	5011.6	20.0	1.73	0.04	-72.0	4.77	4.77	-2.73	0.15	6
196.4117	+14.3176	1.35	0.03	0.55	0.01	5802.4	16.3	4.67	0.01	-50.0	6.3	6.64	-1.06	0.14	5

Continued on next page

Table D.2 – *continued from previous page*

RA_{SDSS}	DEC_{SDSS}	par	dpar	D	dD	T	dT	log(g)	dlg	RV	Fe I	Fe II	$[Fe/H]_{Q6}$	σFe_{Q6}	N,Com
		μas	μas	kpc	kpc	K	K	dex	dex	$km\,s^{-1}$	dex	dex	dex	dex	
196.5453	+12.1211	0.74	0.03	1.28	0.05	6700.0	17.0	4.25	0.03	55.0	5.15	5.15	-2.35	0.11	5
198.5486	+11.4124	0.64	0.03	1.63	0.03	6493.6	46.5	3.83	0.03	77.0	5.06	4.8	-2.55	0.2	5
200.0999	+13.7229	0.01	0.03	20.72	0.45	4775.0	33.0	1.24	0.03	191.0	5.07	4.93	-2.48	0.23	5
200.5298	+08.9768	0.46	0.04	7.14	0.57	5363.3	34.7	2.53	0.08	98.0	6.35	5.77	-1.38	0.21	5
201.1158	+15.4382	1.03	0.03	2.19	0.06	5689.0	66.0	3.37	0.03	-5.0	5.41	5.55	-2.03	0.13	5
201.3732	+08.4513	0.55	0.04	2.80	0.13	5733.0	102.0	3.28	0.06	43.0	6.63	6.39	-0.97	0.08	5
201.8711	+07.1811	0.95	0.04	0.96	0.03	6501.0	27.0	4.44	0.02	32.0	4.55	4.49	-2.97	0.09	4
202.3435	+13.2291	0.46	0.02	2.28	0.09	5749.0	93.0	3.3	0.05	102.0	6.26	6.8	-1.02	0.15	5, D2
203.2831	+13.6326	0.04	0.03	13.13	0.3	5008.0	12.1	1.73	0.03	-140.0	5.17	4.84	-2.44	0.21	6, D2
204.9008	+10.5513	0.66	0.04	1.43	0.07	6717.9	17.4	4.22	0.03	-246.0	4.74	4.62	-2.8	0.19	5
205.1342	+13.8234	0.13	0.05	3.79	0.27	5462.3	28.8	2.9	0.07	127.0	5.68	6.05	-1.63	0.25	4
205.8131	+15.3832	0.74	0.03	1.27	0.05	6718.0	15.0	4.23	0.03	121.0	5.08	5.0	-2.45	0.16	6
206.3487	+09.3099	0.71	0.04	1.49	0.03	6572.0	103.0	3.94	0.03	159.0	5.86	5.52	-1.78	0.2	5
208.0798	+04.4267	0.08	0.04	3.5	0.23	5572.0	66.0	2.97	0.07	-129.0	4.39	4.78	-2.98	0.2	6
209.9364	+15.9251	0.52	0.05	1.93	0.09	6664.0	165.0	3.96	0.06	-92.0	5.14	4.96	-2.42	0.15	6
210.0166	+14.6289	0.16	0.03	5.66	0.31	5287.7	25.5	2.47	0.06	75.0	4.94	4.69	-2.64	0.11	6
210.8632	+08.1798	0.42	0.04	2.68	0.15	5592.0	75.0	3.21	0.06	-12.0	5.39	5.25	-2.16	0.09	6
213.2814	+14.8983	0.20	0.05	2.14	0.03	6001.4	54.4	3.55	0.03	-14.0	6.89	7.26	-0.49	0.1	6
214.5556	+07.4670	0.54	0.03	1.82	0.05	6482.0	203.0	3.88	0.05	47.0	4.93	4.86	-2.59	0.23	6
217.3861	+15.1651	0.78	0.03	1.14	0.04	6663.0	19.0	4.3	0.02	-160.0	6.43	6.1	-1.2	0.13	5
217.5786	+14.0379	0.03	0.03	11.61	0.14	4967.5	12.2	1.64	0.03	-16.0	4.87	4.77	-2.66	0.13	6, D2
217.6443	+15.9634	0.66	0.03	1.32	0.06	6687.0	21.0	4.27	0.03	-20.0	5.92	5.51	-1.72	0.41	3
218.4622	+10.3683	0.08	0.05	2.57	0.04	5923.1	41.2	3.5	0.03	-126.0	5.11	5.75	-2.18	0.41	6
228.4607	+08.3553	0.77	0.04	1.22	0.05	6624.9	25.0	4.32	0.03	9.0	5.09	4.91	-2.47	0.19	6
228.6557	+09.0914	0.70	0.03	1.31	0.06	6695.0	19.8	4.26	0.03	-147.0	5.08	4.98	-2.45	0.15	6

Continued on next page

Table D.2 – *continued from previous page*

RA$_{SDSS}$	DEC$_{SDSS}$	par	dpar	D	dD	T	dT	log(g)	dlg	RV	Fe I	Fe II	[Fe/H]$_{Q6}$	σFe$_{Q6}$	N,Com
		μas	μas	kpc	kpc	K	K	dex	dex	km s^{-1}	dex	dex	dex	dex	
228.8159	+00.2222	0.92	0.06	0.97	0.05	6519.5	32.2	4.41	0.02	6.0	5.52	5.49	-1.99	0.32	6, dw
...	...	...	...	1.72	0.04	6603.2	33.9	3.96	0.03	...	5.61	5.32	-1.98	0.31	6, gi
233.5730	+06.4702	0.10	0.03	11.51	0.29	4990.9	12.7	1.69	0.03	-80.0	4.82	4.38	-2.82	0.26	6
235.1449	+08.7464	0.51	0.03	2.04	0.05	6167.0	87.9	3.64	0.05	-156.0	4.73	4.52	-2.84	0.13	6
236.1068	+10.5311	-0.01	0.03	13.22	0.31	4815.9	15.6	1.47	0.03	-42.0	5.3	5.05	-2.28	0.19	6
237.8246	+10.1427	0.20	0.03	6.01	0.43	5405.0	72.0	2.63	0.08	-165.0	4.76	3.82	-3.05	0.19	6
240.4216	+09.6761	0.11	0.03	8.33	0.41	5204.1	24.1	2.16	0.05	38.0	4.51	4.2	-3.09	0.16	6
245.5747	+06.8844	0.21	0.04	5.31	0.4	5424.0	74.0	2.68	0.08	-189.0	4.69	4.25	-2.98	0.09	5
245.8356	+13.8777	0.15	0.02	4.85	0.23	5257.0	22.2	2.28	0.05	-177.0	4.76	4.4	-2.86	0.19	6
246.5144	+05.9826	1.03	0.03	2.51	0.06	5736.0	53.0	3.28	0.03	-6.0	7.71	7.8	0.25	0.27	5
246.8588	+12.3193	0.19	0.02	7.01	0.34	5172.4	24.2	2.21	0.05	-86.0	5.38	5.12	-2.2	0.22	6
248.4959	+15.0776	0.03	0.02	9.75	0.36	5068.6	19.8	1.86	0.04	-74.0	4.95	5.07	-2.51	0.29	6
250.6963	+08.3743	0.02	0.03	7.82	0.29	5076.2	20.6	2.0	0.04	-4.0	4.99	4.99	-2.51	0.31	6
251.4082	+12.3657	0.08	0.04	14.85	0.43	4918.8	17.4	1.54	0.04	-5.0	4.3	4.0	-3.3	0.11	6
252.1639	+15.0648	0.08	0.04	10.47	0.31	4909.2	18.7	1.65	0.04	-157.0	5.22	5.18	-2.29	0.09	5
253.8582	+15.7240	0.09	0.03	9.17	0.37	5077.0	52.0	2.0	0.05	-97.0	4.92	4.53	-2.71	0.12	6
254.5469	+10.9129	0.41	0.02	2.77	0.16	5517.7	24.3	3.04	0.06	96.0	6.07	5.91	-1.49	0.12	5
255.2671	+14.9712	0.65	0.02	1.51	0.03	6479.0	58.2	3.88	0.04	28.0	5.01	4.9	-2.52	0.04	5
255.5555	+10.8612	0.11	0.03	3.98	0.23	5495.4	23.7	2.86	0.06	-371.0	4.53	5.22	-2.74	0.25	6

Used the SDSS $(g - r)$ solution for [Fe/H]$_P \leq -2.5$:

RA$_{SDSS}$	DEC$_{SDSS}$	par	dpar	D	dD	T	dT	log(g)	dlg	RV	Fe I	Fe II	[Fe/H]$_{Q6}$	σFe$_{Q6}$	N,Com
180.3789	+00.9470	-0.24	0.53	2.02	0.11	5683.5	30.6	4.67	0.01	-58.0	6.05	6.72	-1.18	0.18	5, dw
...	...	...	...	0.40	0.01	5565.0	30.6	3.32	0.01	...	6.45	6.55	-1.00	0.16	5, gi
181.3699	+11.7636	0.74	0.03	2.36	0.12	5494.0	24.8	3.18	0.06	81.0	5.96	5.83	-1.59	0.29	5
188.1264	+08.7740	0.85	0.04	5.88	0.35	5210.5	29.6	2.30	0.07	-49.0	5.55	5.70	-1.89	0.30	5

Continued on next page

Table D.2 – *continued from previous page*

RA$_{SDSS}$	DEC$_{SDSS}$	par	dpar	D	dD	T	dT	log(g)	dlg	RV	Fe I	Fe II	[Fe/H]$_{Q6}$	σFe$_{Q6}$	N,Com
		μas	μas	kpc	kpc	K	K	dex	dex	km s^{-1}	dex	dex	dex	dex	
200.7620	+09.4375	1.01	0.03	0.75	0.01	6237.2	20.3	4.52	0.01	34.0	7.10	6.88	-0.49	0.24	5, dw
...	...	...	...	1.95	0.03	5896.8	49.0	3.61	0.03	...	7.30	6.76	-0.47	0.26	5, gi
229.1219	+00.9089	0.51	0.04	1.84	0.12	6485.3	183.2	3.88	0.12	-222.0	4.91	4.82	-2.62	0.12	6
235.7578	+09.0000	0.60	0.04	1.83	0.07	6653.9	41.1	4.01	0.04	-58.0	5.69	5.51	-1.87	0.04	6
237.9600	+15.4022	0.70	0.03	1.41	0.06	6707.0	15.7	4.09	0.04	-268.0	5.33	5.06	-2.26	0.26	6
245.4387	+08.9954	0.56	0.03	1.87	0.04	6463.6	60.6	3.87	0.04	-83.0	6.07	5.49	-1.62	0.09	6
248.4394	+07.9229	0.95	0.03	0.96	0.02	6636.8	14.6	4.31	0.02	-16.0	5.85	5.65	-1.71	0.17	6
254.5207	+15.4969	1.47	0.02	0.61	0.01	6327.6	11.7	4.49	0.01	25.0	7.30	7.18	-0.25	0.14	5

Low probability for [Fe/H]$_P \leq -2.5$:

RA$_{SDSS}$	DEC$_{SDSS}$	par	dpar	D	dD	T	dT	log(g)	dlg	RV	Fe I	Fe II	[Fe/H]$_{Q6}$	σFe$_{Q6}$	N,Com
182.8512	+14.1595	0.11	0.03	8.88	0.23	4958.7	10.2	1.75	0.02	80.0	5.72	5.95	-1.69	0.17	5
196.6015	+15.6768	1.40	0.05	0.55	0.01	5922.2	24.6	4.62	0.01	-104.0	6.64	6.66	-0.85	0.14	5
197.9861	+12.3578	0.41	0.10	2.23	0.12	6102.0	521.0	3.72	0.13	116.0	6.46	6.15	-1.16	0.06	5
209.2123	+01.5275	0.56	0.04	1.95	0.05	5952.8	95.4	3.57	0.06	13.0	5.81	5.73	-1.72	0.19	6
209.7175	+10.8612	1.23	0.03	0.78	0.02	6358.0	20.0	4.50	0.01	-137.0	5.70	5.41	-1.90	0.14	6
210.7504	+12.7744	0.01	0.04	13.82	0.28	4651.9	15.4	1.35	0.03	41.0	5.33	5.82	-2.01	0.09	6
211.2766	+10.3280	0.79	0.03	1.04	0.03	6527.9	22.1	4.40	0.02	45.0	6.47	6.30	-1.09	0.15	6, dw
...	...	...	...	1.99	0.02	6370.1	39.5	3.81	0.02	...	6.48	6.08	-1.22	0.14	6, gi
212.5834	+10.5365	1.37	0.04	0.66	0.01	6222.0	22.5	4.55	0.01	-127.0	5.68	5.55	-1.86	0.16	6
215.6121	+15.0163	0.21	0.04	3.99	0.22	5267.4	25.2	2.62	0.06	-120.0	6.53	5.92	-1.12	0.10	4
216.1245	+10.2135	0.34	0.04	4.04	0.30	5412.1	31.0	2.78	0.08	107.0	5.26	5.03	-2.31	0.12	6
223.5273	+11.1353	0.00	0.04	10.6	0.25	4757.4	16.4	1.55	0.03	125.0	5.41	5.33	-2.12	0.21	6
227.2885	+01.3598	0.97	0.05	1.79	0.03	5950.4	63.6	3.64	0.04	6.0	7.13	6.82	-0.47	0.25	6
229.0409	+10.3020	1.95	0.04	0.39	0.00	5400.0	16.6	4.71	0.01	-10.0	6.42	6.62	-0.98	0.15	4
232.6956	+08.3392	0.34	0.04	2.96	0.16	5640.6	40.2	3.30	0.06	38.0	5.31	5.01	-2.29	0.24	6

Continued on next page

Table D.2 – *continued from previous page*

RA_{SDSS}	DEC_{SDSS}	par	dpar	D	dD	T	dT	$\log(g)$	dlg	RV	Fe I	Fe II	$[Fe/H]_{Q6}$	σFe_{Q6}	N,Com
		μas	μas	kpc	kpc	K	K	dex	dex	$km\,s^{-1}$	dex	dex	dex	dex	
233.9312	+09.5596	0.38	0.04	2.80	0.14	5645.4	38.5	3.30	0.06	-99.0	5.71	5.10	-2.00	0.09	6
234.4398	+12.6286	1.31	0.04	0.53	0.01	5867.1	22.0	4.63	0.01	-66.0	6.99	6.96	-0.53	0.09	5
235.9710	+09.1864	0.72	0.03	1.54	0.04	6581.5	33.9	4.02	0.03	-32.0	5.91	5.81	-1.62	0.18	6
236.0769	+04.0496	0.78	0.04	1.77	0.04	5753.4	60.2	3.51	0.04	0.0	7.81	...	0.31	0.08	2
241.1186	+09.4156	0.55	0.03	1.76	0.06	6299.3	122.6	3.77	0.07	-52.0	5.65	5.44	-1.92	0.14	6

No longer in the *Pristine* catalogue:

| 213.7879 | +08.4232 | 0.31 | 0.05 | 7.85 | 0.48 | 5289.0 | 29.0 | 2.27 | 0.07 | -107.0 | 4.88 | 4.44 | -2.76 | 0.18 | 6, ** |

Table D.3: Iron-group and heavy element abundances in the new 28 metal-poor stars, and our analysis of HD122563. [Fe/H] is calculated as the weighted mean of Fe I and Fe II, and $\delta([Fe/H]) = \sigma([Fe/H])/(NFe_I + NFe_{II}))^{1/2}$.

RA_{SDSS}	[Fe/H] $\pm\delta$	log(Fe I) $\pm\sigma$ (N)	log(Fe II) $\pm\sigma$ (N)	[Cr/Fe] $\pm\sigma$ (N)	[Ni/Fe] $\pm\sigma$ (N)	[Y/Fe] $\pm\sigma$ (N)	[Ba/Fe] $\pm\sigma$ (N)
180.2206	−2.92 ± 0.03	4.60 ±0.18 (49)	4.45 ±0.27 (4)	−0.21 ± 0.13 (2)	−0.05 ± 0.18 (1)	< +0.49	−1.02 ± 0.18 (1)
181.2243	−3.11 ± 0.07	4.50 ±0.17 (4)	4.17 ±0.14 (2)	...	...	< +1.85	< +1.37
181.4395	−2.52 ± 0.02	4.99 ±0.18 (86)	4.80 ±0.26 (5)	−0.37 ± 0.11 (3)	+0.03 ± 0.13 (2)	< −0.12	−1.13 ± 0.24 (2)
183.6849	−3.16 ± 0.07	4.38 ±0.24 (2)	4.30 ±0.01 (2)	...	...	< +2.33	< +1.88
189.9449	−2.78 ± 0.04	4.77 ±0.12 (8)	4.50 ±0.21 (2)	...	...	< +1.70	< +0.93
193.8390	−2.80 ± 0.02	4.71 ±0.21 (81)	4.49 ±0.13 (5)	−0.49 ± 0.12 (3)	−0.24 ± 0.21 (1)	< +0.02	−1.58 ± 0.21 (1)
196.3755	−2.80 ± 0.02	4.70 ±0.19 (61)	4.72 ±0.16 (3)	−0.17 ± 0.14 (2)	−0.21 ± 0.19 (1)	< +0.3	< −0.56
198.5486	−2.47 ± 0.05	5.08 ±0.16 (9)	4.80 ±0.29 (2)	+0.04 ± 0.11 (2)	< +0.45	< +1.23	< +0.39
201.8711	−2.93 ± 0.11	4.57 ±0.15 (2)	< 4.71	...	...	< +2.10	< +1.65
203.2831	−2.74 ± 0.02	4.77 ±0.19 (59)	4.55 ±0.15 (4)	−0.27 ± 0.11 (3)	−0.19 ± 0.19 (1)	< +0.01	−0.76 ± 0.23 (3)
204.9008	−2.73 ± 0.08	4.84 ±0.18 (3)	4.66 ±0.26 (2)	...	...	< +1.89	< +1.40
208.0798	−2.94 ± 0.03	4.53 ±0.14 (20)	4.83 ±0.26 (2)	−0.35 ± 0.10 (2)	< +0.16	< +1.15	< +0.57
210.0166	−2.59 ± 0.02	4.92 ±0.18 (64)	4.75 ±0.13 (4)	−0.20 ± 0.13 (2)	−0.17 ± 0.18 (1)	< +0.51	+0.74 ± 0.20 (4)
213.7879	−2.59 ± 0.02	4.93 ±0.18 (61)	4.55 ±0.17 (3)	−0.09 ± 0.11 (3)	+0.11 ± 0.18 (1)	< +0.43	−0.36 ± 0.14 (2)
214.5556	−2.51 ± 0.05	5.01 ±0.15 (10)	4.88 ±0.35 (2)	< +0.07	< +0.69	+1.48 ± 0.11 (2)	+1.77 ± 0.22 (4)
217.5786	−2.61 ± 0.02	4.88 ±0.18 (79)	4.94 ±0.18 (6)	−0.11 ± 0.16 (4)	−0.04 ± 0.18 (1)	+0.16 ± 0.13 (2)	+0.02 ± 0.18 (3)
229.1219	−2.52 ± 0.04	5.02 ±0.12 (8)	4.84 ±0.17 (2)	+0.16 ± 0.12 (1)	...	< +1.59	< +0.44
233.5730	−2.75 ± 0.02	4.77 ±0.20 (75)	4.42 ±0.28 (3)	−0.25 ± 0.12 (3)	...	< +0.23	−0.15 ± 0.32 (3)
235.1449	−2.69 ± 0.04	4.85 ±0.15 (16)	4.52 ±0.13 (2)	< −0.16	...	< +1.41	+0.13 ± 0.15 (1)
237.8246	−3.29 ± 0.04	4.28 ±0.15 (13)	3.73 ±0.04 (2)	< −0.32	< +0.50	< +1.29	+0.61 ± 0.34 (2)
240.4216	−2.95 ± 0.03	4.56 ±0.20 (41)	4.20 ±0.22 (2)	−0.35 ± 0.14 (2)	−0.19 ± 0.20 (1)	< +0.64	+0.53 ± 0.29 (4)
245.5747	−3.14 ± 0.04	4.37 ±0.20 (18)	4.25 ±0.09 (2)	< −0.30	< +0.38	< +1.14	< +0.52
245.8356	−2.78 ± 0.03	4.73 ±0.21 (52)	4.60 ±0.25 (4)	−0.41 ± 0.16 (5)	+0.18 ± 0.21 (1)	+0.73 ± 0.21 (1)	−0.51 ± 0.12 (3)
248.4959	−2.59 ± 0.02	4.91 ±0.17 (74)	4.90 ±0.31 (5)	−0.26 ± 0.10 (3)	−0.04 ± 0.12 (2)	< +0.16	−0.18 ± 0.10 (3)

Continued on next page

Table D.3 – *continued from previous page*

RA$_{\text{SDSS}}$	[Fe/H] $\pm\delta$	log(Fe I) $\pm\sigma$ (N)	log(Fe II) $\pm\sigma$ (N)	[Cr/Fe] $\pm\sigma$ (N)	[Ni/Fe] $\pm\sigma$ (N)	[Y/Fe] $\pm\sigma$ (N)	[Ba/Fe] $\pm\sigma$ (N)
250.6963	-2.55 ± 0.03	4.95 $\pm$0.20 (62)	4.95 $\pm$0.28 (4)	-0.43 ± 0.12 (3)	$+0.07 \pm 0.14$ (2)	< +0.18	-0.03 ± 0.16 (3)
251.4082	-3.27 ± 0.03	4.23 $\pm$0.18 (31)	4.19 $\pm$0.33 (3)	< -0.17	...	< +0.65	-0.73 ± 0.18 (1)
253.8582	-2.72 ± 0.02	4.80 $\pm$0.21 (65)	4.54 $\pm$0.07 (4)	-0.15 ± 0.14 (2)	-0.13 ± 0.21 (1)	< +0.35	-0.51 ± 0.20 (3)
255.5555	-2.83 ± 0.03	4.64 $\pm$0.18 (22)	4.99 $\pm$0.10 (2)	< -0.08	< -0.08	< +0.97	$+0.35 \pm 0.10$ (3)
HD122563	-2.76 ± 0.01	4.73 $\pm$0.15 (98)	4.86 $\pm$0.16 (5)	-0.39 ± 0.08 (3)	$+0.05 \pm 0.10$ (5)	-0.15 ± 0.11 (2)	-0.77 ± 0.08 (3)

Table D.4: Light element abundances in the new 28 metal-poor stars (and our analysis of HD122563). When the number of lines $N_X < 4$ for species X, then $\sigma(X) = \sigma\mathrm{Fe\,I}/\sqrt{N_X}$.

RA$_{\mathrm{SDSS}}$	[Na/Fe] $\pm\sigma$ (N)	[Mg/Fe] $\pm\sigma$ (N)	[Ca/Fe] $\pm\sigma$ (N)	[Sc/Fe] $\pm\sigma$ (N)	[TiI/Fe] $\pm\sigma$ (N)	[TiII/Fe] $\pm\sigma$ (N)
180.2206	$+0.39 \pm 0.13$ (2)	$+0.30 \pm 0.11$ (3)	$+0.35 \pm 0.10$ (4)	...	$+0.22 \pm 0.13$ (2)	$+0.28 \pm 0.11$ (3)
181.2243	-0.23 ± 0.17 (1)	$+0.67 \pm 0.13$ (2)	$< +0.94$	...	...	$< +1.23$
181.4395	$+0.07 \pm 0.13$ (2)	$+0.26 \pm 0.11$ (3)	$+0.42 \pm 0.19$ (9)	-0.15 ± 0.11 (3)	$+0.16 \pm 0.19$ (9)	$+0.19 \pm 0.16$ (8)
183.6849	-0.18 ± 0.17 (2)	$+0.13 \pm 0.14$ (3)	$< +1.01$	...	...	$< +1.88$
189.9449	$+0.01 \pm 0.08$ (2)	$+0.17 \pm 0.08$ (2)	$< +0.65$	...	...	$< +0.92$
193.8390	$< +1.17$	$+0.37 \pm 0.12$ (3)	$+0.33 \pm 0.14$ (8)	$+0.08 \pm 0.12$ (3)	$+0.09 \pm 0.12$ (8)	$+0.17 \pm 0.18$ (8)
196.3755	$+0.22 \pm 0.13$ (2)	$+0.29 \pm 0.11$ (3)	$+0.45 \pm 0.24$ (7)	...	$+0.43 \pm 0.25$ (4)	$+0.28 \pm 0.19$ (4)
198.5486	$+0.24 \pm 0.11$ (2)	$+0.25 \pm 0.09$ (3)	$+0.53 \pm 0.16$ (1)	$< +0.97$	...	$< +0.71$
201.8711	-0.12 ± 0.11 (2)	-0.05 ± 0.11 (2)	$< +1.16$	...	...	$< +1.59$
203.2831	$+0.66 \pm 0.13$ (2)	$+0.02 \pm 0.13$ (2)	$+0.21 \pm 0.11$ (3)	...	$+0.43 \pm 0.06$ (6)	$+0.50 \pm 0.10$ (5)
204.9008	-0.15 ± 0.18 (1)	-0.12 ± 0.13 (2)	$< +0.96$	...	...	$< +1.36$
208.0798	$+0.26 \pm 0.10$ (2)	$+0.09 \pm 0.08$ (3)	$+0.45 \pm 0.14$ (1)	...	...	$< +0.78$
210.0166	$+0.06 \pm 0.13$ (2)	$+0.18 \pm 0.11$ (3)	$+0.36 \pm 0.28$ (4)	$< +0.44$	$+0.37 \pm 0.11$ (3)	$+0.12 \pm 0.13$ (2)
213.7879	-0.12 ± 0.13 (2)	$+0.33 \pm 0.11$ (3)	$+0.44 \pm 0.09$ (4)	$< +0.26$	$+0.45 \pm 0.24$ (7)	$+0.45 \pm 0.13$ (2)
214.5556	$+0.02 \pm 0.12$ (2)	$+0.36 \pm 0.09$ (3)	$< +0.82$	$< +1.11$	...	$< +0.96$
217.5786	$< +1.14$	$+0.22 \pm 0.11$ (3)	$+0.55 \pm 0.24$ (6)	$+0.09 \pm 0.13$ (2)	$+0.23 \pm 0.12$ (5)	$+0.55 \pm 0.16$ (10)
229.1219	-0.13 ± 0.09 (2)	$+0.22 \pm 0.09$ (2)	....	...	$+1.26 \pm 0.12$ (1)	$+1.44 \pm 0.12$ (1)
233.5730	$+0.65 \pm 0.14$ (2)	$+0.24 \pm 0.12$ (3)	$+0.39 \pm 0.32$ (5)	$< +0.10$	$+0.24 \pm 0.15$ (4)	$+0.28 \pm 0.12$ (3)
235.1449	-0.26 ± 0.11 (2)	$+0.09 \pm 0.11$ (2)	$< +0.67$	$< +1.16$	...	$< +0.94$
237.8246	$+0.00 \pm 0.11$ (2)	$+0.04 \pm 0.11$ (2)	$< +0.88$	$< +1.12$	...	$< +0.89$

Continued on next page

Table D.4 – *continued from previous page*

RA_{SDSS}	[Na/Fe] $\pm\sigma$ (N)	[Mg/Fe] $\pm\sigma$ (N)	[Ca/Fe] $\pm\sigma$ (N)	[Sc/Fe] $\pm\sigma$ (N)	[TiI/Fe] $\pm\sigma$ (N)	[TiII/Fe] $\pm\sigma$ (N)
240.4216	$+0.14 \pm 0.14$ (2)	$+0.14 \pm 0.12$ (3)	$+0.31 \pm 0.12$ (3)	$< +0.54$	$+0.44 \pm 0.14$ (2)	$+0.52 \pm 0.14$ (2)
245.5747	-0.17 ± 0.14 (2)	$+0.14 \pm 0.14$ (2)	$< +0.96$	$< +0.99$	...	$< +0.75$
245.8356	$+0.67 \pm 0.15$ (2)	$+0.29 \pm 0.11$ (4)	$+0.66 \pm 0.22$ (6)	$+0.30 \pm 0.15$ (2)	$+0.64 \pm 0.15$ (2)	$+0.46 \pm 0.25$ (11)
248.4959	$+0.39 \pm 0.13$ (2)	$+0.14 \pm 0.10$ (3)	$+0.35 \pm 0.17$ (7)	$< +0.07$	$+0.48 \pm 0.28$ (4)	$+0.36 \pm 0.18$ (5)
250.6963	$< +1.71$	$+0.11 \pm 0.12$ (3)	$+0.57 \pm 0.38$ (8)	$+0.03 \pm 0.20$ (1)	$+0.40 \pm 0.24$ (6)	$+0.39 \pm 0.12$ (8)
251.4082	$< +1.61$	-0.20 ± 0.13 (2)	$< +0.11$	$< +0.52$	...	$< +0.28$
253.8582	$+0.12 \pm 0.15$ (2)	$+0.23 \pm 0.12$ (3)	$+0.27 \pm 0.20$ (4)	$< +0.20$	$+0.39 \pm 0.24$ (4)	$+0.37 \pm 0.18$ (5)
255.5555	$+0.76 \pm 0.13$ (2)	$+0.29 \pm 0.11$ (3)	$+0.46 \pm 0.10$ (3)	$< +0.78$	...	$< +0.88$
HD122563	$+0.21 \pm 0.11$ (2)	$+0.18 \pm 0.09$ (3)	$+0.32 \pm 0.14$ (8)	$+0.19 \pm 0.11$ (2)	$+0.07 \pm 0.05$ (9)	$+0.46 \pm 0.08$ (9)

Table D.5: Total systematic errors (dX) per element species (X) due to the stellar parameters (T, logg, [Fe/H]) added in quadrature, where dT and dlogg are from Table D.2 and δ[Fe/H] is from Table D.3.

RA$_{SDSS}$	dFeI	dFeII	dNa	dMg	dCa	dSc	dTiI	dTiII	dCrI	dNiI	dYII	dBaII
180.2206	0.03	0.02	0.03	0.02	0.01	...	0.02	0.02	0.02	0.02	...	0.02
181.2243	0.09	0.03	0.07	0.08	...	...	...	...	...	...	...	...
181.4395	0.01	0.01	0.01	0.01	0.01	0.01	0.01	0.01	0.01	0.01	...	...
183.6849	0.04	0.01	0.03	0.03	...	...	...	...	...	...	...	...
189.9449	0.12	0.02	0.09	0.09	...	...	...	...	...	...	...	...
193.8390	0.04	0.01	0.05	0.04	0.03	0.01	0.05	0.01	0.04	0.04	...	...
196.3755	0.02	0.01	0.02	0.02	0.01	...	0.03	0.01	0.02	0.02	...	...
198.5486	0.04	0.01	0.03	0.03	0.03	...	...	...	0.04	...	...	...
201.8711	0.03	...	0.02	0.02	...	...	...	...	...	...	...	...
203.2831	0.01	0.01	0.02	0.02	0.01	...	0.02	0.01	0.01	0.01	...	0.01
204.9008	0.01	0.01	0.01	0.01	...	...	...	...	...	...	...	...
208.0798	0.07	0.03	0.07	0.06	0.05	...	...	...	0.06	...	...	...
210.0166	0.03	0.02	0.03	0.03	0.02	...	0.03	0.02	0.03	0.03	...	0.03
213.7879	0.03	0.03	0.03	0.04	0.02	...	0.03	0.03	0.03	0.03	...	0.03
214.5556	0.17	0.03	0.13	0.14	...	...	...	...	...	...	0.09	0.15
217.5786	0.01	0.01	0.02	0.02	0.01	0.01	0.02	0.01	0.01	0.01	...	...
229.1219	0.16	0.05	0.12	0.15	...	...	0.15	0.06	0.15	...	0.08	...
233.5730	0.02	0.01	0.02	0.02	0.01	...	0.02	0.01	0.02	...	...	0.01
235.1449	0.08	0.02	0.06	0.08	...	...	...	...	...	...	...	0.06
237.8246	0.09	0.03	0.07	0.07	...	...	...	...	...	...	...	0.06

Continued on next page

Table D.5 – *continued from previous page*

RA$_{SDSS}$	dFeI	dFeII	dNa	dMg	dCa	dSc	dTiI	dTiII	dCrI	dNiI	dYII	dBaII
240.4216	0.03	0.02	0.02	0.02	0.02	...	0.03	0.02	0.03	0.03	...	0.02
245.5747	0.08	0.03	0.07	0.08	...	...	...	...	...	...	...	...
245.8356	0.02	0.02	0.03	0.02	0.02	0.02	0.03	0.02	0.03	0.02	...	...
248.4959	0.02	0.01	0.03	0.02	0.01	...	0.02	0.02	0.02	0.02	...	0.02
250.6963	0.02	0.01	0.04	0.02	0.02	0.02	0.02	0.02	0.02	0.02	...	...
251.4082	0.02	0.01	0.03	0.02	...	...	...	...	...	...	...	0.02
253.8582	0.06	0.02	0.06	0.06	0.04	...	0.06	0.03	0.06	0.06	...	0.04
255.5555	0.03	0.02	0.03	0.03	0.02	...	...	...	...	...	...	0.02
HD122563	0.03	0.02	0.04	0.03	0.02	0.01	0.04	0.01	0.03	0.03	...	...

Table D.6: Samples of the systematic errors from the individual stellar parameters per elemental species. These were added in quadrature for these stars (and the rest of the sample) in Table D.5. We note that if σFe I is used instead of σ[Fe/H], then the errors due to metallicity remain negligible.

RA$_{SDSS}$	Param ±d	dFe I	dFe II	dNa I	dMg I	dCa I	dSc II	dTi I	dTi II	dCr I	dNi I	dY II	dBa II
193.8390	4764 ±32	0.04	−0.01	0.05	0.03	0.03	0.01	0.05	0.01	0.04	0.04	...	...
193.8390	1.22 ±0.03	0.00	0.01	−0.01	−0.01	0.00	0.01	0.00	0.01	0.00	0.00	...	...
193.8390	−2.80±0.02	0.00	0.00	0.00	0.00	0.00	0.00	0.00	0.00	0.00	0.00	...	...
213.7879	5289 ±29	0.03	0.00	0.03	0.03	0.02	...	0.03	0.01	0.03	0.03	...	0.02
213.7879	2.27 ±0.07	0.00	0.03	0.00	−0.02	0.00	...	0.00	0.02	0.00	0.00	...	0.02
213.7879	−2.59±0.02	0.00	0.00	0.00	0.00	0.00	...	0.00	0.00	0.00	0.00	...	0.00
214.5556	6482 ±203	0.17	0.02	0.13	0.14	...	...	...	...	...	...	0.09	0.15
214.5556	3.88 ±0.05	0.00	0.02	0.00	−0.01	...	...	...	...	...	...	0.02	0.01
214.5556	−2.50±0.05	0.00	0.00	0.00	0.00	...	...	...	...	...	...	0.00	0.00

Table D.7: Orbit and Action parameters for the 70 metal-poor candidates from the *Pristine* survey.

RA	DEC	pmra (μas yr^{-1})	pmdec (μas yr^{-1})	U (km s^{-1})	V (km s^{-1})	W (km s^{-1})	R_{apo} (kpc)	R_{peri} (kpc)	ecc	Z_{max} (kpc)	$J_\phi/J_{\phi\odot}$	$J_z/J_{z\odot}$	$E/E_\odot$
180.2206	9.5683	1.47 ± 0.08	-3.68 ± 0.06	175.69± 6.90	-143.29± 5.31	-26.54± 1.96	23.5 ± 1.3	6.6 ± 0.5	0.57 ± 0.01	23.4 ± 1.3	-0.04± 0.05	5918.45± 289.73	0.477± -0.032
180.3789	0.9470	-14.76 ± 0.92	-5.28 ± 0.49	-101.97± 6.13	-73.77± 5.78	-100.36± 2.89	9.4 ± 0.3	6.3 ± 0.2	0.20 ± 0.03	4.5 ± 0.4	0.67± 0.03	689.95± 82.44	0.893± -0.011
181.2243	7.4160	1.98 ± 0.08	-8.19 ± 0.05	60.40± 1.88	-4.87± 1.86	-161.93± 0.92	19.0 ± 0.1	6.5 ± 0.1	0.49 ± 0.01	9.4 ± 0.1	0.87± 0.01	948.17± 13.67	0.58± -0.003
181.3464	11.6698	-14.31 ± 0.06	2.84 ± 0.03	-39.73± 0.65	-15.73± 0.26	6.31± 0.48	8.8 ± 0.0	7.5 ± 0.1	0.08 ± 0.00	0.7 ± 0.1	0.87± 0.0	32.49± 0.77	0.881± -0.001
181.3699	11.7636	-9.20 ± 0.06	-4.69 ± 0.03	-66.05± 3.11	-116.49± 4.39	44.36± 1.64	8.9 ± 0.1	3.4 ± 0.2	0.45 ± 0.02	2.9 ± 0.1	0.48± 0.02	303.79± 5.57	1.045± -0.002
181.4395	1.6294	-3.34 ± 0.08	-9.14 ± 0.04	134.17± 3.39	-768.5± 11.62	-177.4± 6.21	2345.5 ± 124.2	18.1 ± 0.3	0.98 ± 0.00	1049.8 ± 44.4	-2.4± 0.05	16291.42± 799.81	-1.919± -0.125
181.6953	13.8075	-16.18 ± 0.06	-3.44 ± 0.03	-164.22± 6.23	-156.8± 5.18	28.91± 1.87	11.7 ± 0.3	2.1 ± 0.1	0.70 ± 0.02	3.7 ± 0.2	0.36± 0.02	271.81± 17.54	0.94± -0.012
182.5364	0.9431	-43.19 ± 0.08	-49.25 ± 0.06	-124.30± 7.39	-702.83± 30.92	-103.95± 15.46	541.2 ± 437.2	7.9 ± 0.3	0.94 ± 0.04	164.9 ± 134.9	-1.77± 0.12	766.27± 167.98	-0.66± -0.285
183.6849	4.8619	-0.70 ± 0.11	-12.87 ± 0.12	31.50± 1.67	-68.32± 2.46	11.5± 1.24	8.5 ± 0.1	4.5 ± 0.1	0.30 ± 0.01	1.2 ± 0.1	0.64± 0.01	84.71± 4.01	1.022± -0.005
185.4110	7.4777	-70.21 ± 0.07	-37.42 ± 0.04	-146.72± 3.89	-305.15± 5.95	85.86± 1.97	10.7 ± 0.2	1.5 ± 0.1	0.75 ± 0.01	4.0 ± 0.2	-0.27± 0.02	352.7± 24.77	1.002± -0.012
187.9785	8.7294	-20.24 ± 0.10	2.84 ± 0.05	121.345± 6.39	-32.49± 2.74	-55.94± 0.52	11.5 ± 0.2	6.0 ± 0.1	0.32 ± 0.02	2.9 ± 0.2	0.80± 0.01	240.41± 19.72	0.823± 0.004
188.1264	8.7740	-6.31 ± 0.08	-7.70 ± 0.05	-47.33± 2.51	-247.10± 14.74	-126.96± 4.56	9.9 ± 0.2	3.2 ± 0.6	0.51 ± 0.06	9.8 ± 0.2	-0.02± 0.05	2933.74± 261.61	0.96± -0.028
189.9449	11.5534	3.40 ± 0.06	-55.97 ± 0.05	257.42± 4.44	-358.77± 6.12	-88.47± 2.14	24.8 ± 1.4	2.7 ± 0.1	0.81 ± 0.00	5.2 ± 0.4	-0.55± 0.03	194.46± 6.74	0.502± -0.034
190.6313	8.5137	-13.53 ± 0.10	-19.24 ± 0.05	-34.76± 1.66	-330.56± 21.07	-163.11± 6.21	8.7 ± 0.4	5.9 ± 1.2	0.20 ± 0.09	7.4 ± 0.3	-0.36± 0.08	2128.14± 48.96	0.924± -0.057
193.1159	8.0557	6.70 ± 0.06	-57.85 ± 0.04	140.26± 3.27	-169.78± 3.91	-30.26± 1.81	10.5 ± 0.1	1.1 ± 0.1	0.81 ± 0.02	1.0 ± 0.1	0.23± 0.02	44.76± 2.42	1.048± -0.005
193.5533	11.5037	1.37 ± 0.05	-1.21 ± 0.03	7.22± 0.13	-4.11± 0.12	14.51± 0.48	9.3 ± 0.0	7.7 ± 0.0	0.10 ± 0.00	0.8 ± 0.1	0.9± 0.0	40.68± 1.03	0.852± -0.0
193.8390	11.4150	0.98 ± 0.07	-2.19 ± 0.04	178.73± 4.59	-112.05± 3.29	-51.18± 1.35	33.6 ± 2.0	13.8 ± 0.5	0.42 ± 0.01	33.5 ± 1.9	-0.1± 0.03	10599.81± 378.3	0.235± -0.029
196.3755	8.5138	-1.46 ± 0.08	-6.48 ± 0.06	106.90± 4.63	-317.24± 12.03	-179.96± 4.05	23.9 ± 2.2	4.9 ± 0.8	0.66 ± 0.02	21.1 ± 2.1	-0.39± 0.04	2991.74± 412.01	0.49± -0.054
196.4117	14.3176	-18.78 ± 0.06	-9.71 ± 0.04	-35.04± 0.48	-40.59± 0.87	-51.6± 0.49	8.1 ± 0.1	6.4 ± 0.2	0.12 ± 0.00	1.2 ± 0.1	0.75± 0.0	86.87± 1.97	0.958± -0.002
196.5453	12.1211	25.63 ± 0.07	-56.26 ± 0.04	321.48± 11.69	-196.02± 6.99	-44.4± 3.69	31.8 ± 3.5	0.5 ± 0.1	0.97 ± 0.01	4.0 ± 0.7	0.1± 0.03	96.63± 5.86	0.38± -0.061
198.5486	11.4123	-47.39 ± 0.06	-46.31 ± 0.04	-86.24± 1.88	-510.48± 8.99	8.85± 1.27	17.6 ± 1.6	7.1 ± 0.1	0.42 ± 0.03	3.2 ± 0.4	-1.04± 0.03	174.96± 8.22	0.61± -0.041
200.0999	13.7228	-4.09 ± 0.06	-1.02 ± 0.04	-223.20± 6.01	-340.64± 7.01	206.07± 0.83	90.2 ± 10.1	9.9 ± 0.7	0.80 ± 0.01	89.2 ± 9.5	0.10± 0.01	10765.9± 780.77	-0.176± -0.048
200.5298	8.9768	-20.91 ± 0.08	-12.61 ± 0.04	-69.46± 8.35	-233.94± 18.62	78.19± 1.31	12.9 ± 51.3	0.4 ± 0.2	0.92 ± 0.06	5.8 ± 19.7	0.00± 0.06	472.31± 14.80	1.17± 0.01
200.7620	9.4375	8.03 ± 0.06	-4.98 ± 0.03	93.28± 1.30	1.62± 0.24	8.97± 0.59	13.6 ± 0.1	6.0 ± 0.1	0.39 ± 0.00	3.6 ± 0.1	0.85± 0.0	273.96± 7.04	0.748± -0.003
201.1158	15.4381	-8.64 ± 0.06	-1.74 ± 0.05	-62.21± 1.65	-67.21± 1.82	2.9± 0.53	8.3 ± 0.0	4.8 ± 0.1	0.27 ± 0.01	2.3 ± 0.1	0.62± 0.01	240.53± 10.05	1.01± -0.002
201.3732	8.4513	-17.04 ± 0.07	-3.15 ± 0.04	-144.52± 6.99	-175.08± 7.42	56.68± 0.87	10.1 ± 0.3	1.4 ± 0.2	0.76 ± 0.04	3.6 ± 0.2	0.24± 0.03	321.08± 11.49	1.049± -0.012
201.8711	7.1810	-78.20 ± 0.08	-13.09 ± 0.05	-239.77± 7.57	-265.05± 7.84	58.94± 1.00	15.0 ± 0.8	0.5 ± 0.1	0.94 ± 0.01	3.2 ± 1.0	-0.1± 0.03	124.81± 1.95	0.822± -0.032
202.3435	13.2291	-4.59 ± 0.05	-26.08 ± 0.02	145.13± 4.50	-264.49± 9.62	36.58± 2.37	10.1 ± 0.2	1.0 ± 0.2	0.83 ± 0.04	6.0 ± 0.2	-0.14± 0.04	640.0± 15.69	1.035± -0.016
203.2831	13.6326	-0.91 ± 0.05	-5.35 ± 0.03	105.34± 3.35	-284.55± 6.72	-203.82± 1.65	27.6 ± 1.3	2.9 ± 0.4	0.81 ± 0.02	25.5 ± 1.7	-0.19± 0.02	2748.57± 254.75	0.432± -0.028
204.9008	10.5513	10.65 ± 0.06	-33.88 ± 0.05	103.14± 8.51	-106.04± 6.33	-310.79± 3.75	31.5 ± 1.5	6.2 ± 0.2	0.67 ± 0.02	28.3 ± 1.6	0.47± 0.03	3873.18± 104.35	0.336± -0.022
205.1342	13.8234	-8.1 ± 0.07	-2.72 ± 0.06	-46.57± 5.60	-136.76± 8.5	137.12± 1.18	9.5 ± 0.3	3.6 ± 0.4	0.45 ± 0.05	7.2 ± 0.4	0.33± 0.03	1413.11± 31.15	0.976± -0.003
205.8131	15.3832	10.45 ± 0.06	-30.98 ± 0.04	190.87± 5.92	-113.35± 4.09	64.18± 2.01	14.3 ± 0.3	2.8 ± 0.1	0.68 ± 0.02	6.3 ± 0.3	0.45± 0.02	541.15± 12.38	0.801± -0.01
206.3487	9.3099	1.22 ± 0.06	-67.65 ± 0.04	324.56± 5.30	-385.14± 7.22	-2.39± 3.00	41.4 ± 3.5	2.7 ± 0.1	0.88 ± 0.01	10.2 ± 0.8	-0.6± 0.03	246.21± 2.21	0.223± -0.045
208.0798	4.4266	2.02 ± 0.07	-12.87 ± 0.07	81.27± 8.37	-120.10± 8.76	-205.75± 5.62	13.8 ± 0.7	3.4 ± 0.4	0.61 ± 0.05	10.7 ± 0.9	0.34± 0.04	1600.13± 53.45	0.8± -0.02
209.9364	15.9251	1.78 ± 0.08	-13.73 ± 0.07	56.47± 3.89	-87.63± 3.89	-116.8± 1.42	9.3 ± 0.1	4.2 ± 0.2	0.37 ± 0.02	4.3 ± 0.1	0.53± 0.02	578.82± 17.2	0.976± -0.002
210.0166	14.6289	-9.65 ± 0.05	-10.23 ± 0.05	-0.72± 1.52	-377.07± 19.55	76.2± 0.60	8.8 ± 0.5	5.1 ± 0.9	0.27 ± 0.06	6.4 ± 0.4	-0.43± 0.05	1421.89± 143.6	0.959± -0.057
210.8632	8.1797	-5.11 ± 0.06	-13.39 ± 0.05	41.58± 2.50	-170.72± 9.11	-49.28± 2.09	7.8 ± 0.1	1.4 ± 0.3	0.69 ± 0.05	2.8 ± 0.1	0.20± 0.04	300.71± 22.72	1.214± -0.004
213.2814	14.8983	-0.27 ± 0.08	-13.33 ± 0.07	74.25± 1.18	-105.65± 1.51	-42.69± 0.64	8.6 ± 0.1	3.0 ± 0.1	0.48 ± 0.01	2.3 ± 0.1	0.45± 0.01	220.76± 4.28	1.084± -0.002
213.7879	8.4232	5.67 ± 0.07	-15.94 ± 0.07	132.56± 27.58	-114.70± 18.31	-194.86± 15.19	15.8 ± 2.8	3.0 ± 0.7	0.67 ± 0.11	10.8 ± 2.4	0.38± 0.08	1174.7± 64.97	0.738± 0.077
214.5556	7.4670	-43.60 ± 0.05	-28.18 ± 0.05	-102.93± 3.36	-430.78± 11.45	81.29± 1.16	11.2 ± 0.8	5.1 ± 0.2	0.37 ± 0.01	3.5 ± 0.2	-0.70± 0.04	340.16± 6.67	0.864± -0.042
217.3861	15.1650	6.02 ± 0.06	-12.35 ± 0.06	-5.72± 2.14	-41.35± 1.0	-171.37± 1.03	10.9 ± 0.1	7.5 ± 0.1	0.19 ± 0.00	6.4 ± 0.1	0.71± 0.01	1150.58± 17.37	0.793± -0.002
217.5786	14.0379	-1.42 ± 0.05	-7.46 ± 0.04	191.61± 2.64	-361.55± 4.47	-86.38± 1.13	22.5 ± 1.0	7.5 ± 0.2	0.50 ± 0.01	22.4 ± 1.0	-0.11± 0.0	6103.43± 214.15	0.483± -0.024
217.6443	15.9633	-1.04 ± 0.06	-16.84 ± 0.06	53.01± 2.68	-85.20± 3.63	-38.39± 1.00	8.2 ± 0.0	3.7 ± 0.1	0.38 ± 0.02	1.5 ± 0.1	0.55± 0.02	112.75± 7.09	1.076± -0.006
218.4622	10.3683	-2.14 ± 0.07	-8.22 ± 0.07	-20.40± 0.78	-95.99± 1.48	-130.24± 0.58	7.4 ± 0.0	4.9 ± 0.1	0.20 ± 0.01	4.6 ± 0.1	0.46± 0.01	968.5± 12.36	1.038± -0.003

Continued on next page

Table D.7 – *continued from previous page*

RA	DEC	pmra (μas yr^{-1})	pmdec (μas yr^{-1})	U (km s^{-1})	V (km s^{-1})	W (km s^{-1})	R_{apo} (kpc)	R_{peri} (kpc)	ecc	Z_{max} (kpc)	$J_\phi/J_{\phi\odot}$	$J_z/J_{z\odot}$	$E/E_\odot$
228.4607	8.3553	-3.84 ± 0.07	-59.49 ± 0.09	192.77 ± 7.38	-267.75 ± 10.58	-100.5 ± 4.25	12.9 ± 0.7	0.9 ± 0.1	0.87 ± 0.01	4.3 ± 0.2	-0.11 ± 0.04	261.09 ± 20.8	0.909 ± -0.036
228.6557	9.0914	-15.11 ± 0.05	-33.13 ± 0.06	-18.70 ± 3.11	-232.36 ± 9.37	-135.68 ± 0.98	7.5 ± 0.0	0.6 ± 0.2	0.86 ± 0.05	7.2 ± 0.3	-0.0 ± 0.04	1206.47 ± 29.75	1.181 ± -0.002
228.8159	0.2222	-6.67 ± 0.09	-59.12 ± 0.11	120.03 ± 4.60	-223.37 ± 8.85	-102.83 ± 4.26	9.7 ± 0.4	0.4 ± 0.2	0.92 ± 0.04	4.1 ± 0.3	0.03 ± 0.03	349.16 ± 37.54	1.104 ± -0.014
229.1219	0.9089	-31.52 ± 0.06	-33.97 ± 0.05	-151.99 ± 0.39	-409.54 ± 24.74	-147.89 ± 0.89	12.9 ± 1.6	5.2 ± 0.4	0.43 ± 0.02	7.8 ± 0.5	-0.59 ± 0.08	1122.98 ± 59.34	0.79 ± -0.064
233.5730	6.4702	-3.73 ± 0.04	-6.44 ± 0.05	47.28 ± 2.63	-404.88 ± 9.71	-71.97 ± 0.93	11.6 ± 0.8	5.1 ± 0.5	0.40 ± 0.02	11.6 ± 0.8	0.03 ± 0.02	4111.92 ± 239.13	0.83 ± -0.044
235.1449	8.7463	16.68 ± 0.05	-25.74 ± 0.05	116.75 ± 5.29	-104.67 ± 1.79	-296.21 ± 4.42	25.6 ± 1.3	5.2 ± 0.1	0.66 ± 0.02	22.0 ± 1.3	0.45 ± 0.01	2973.52 ± 51.91	0.45 ± -0.026
235.7578	9.0000	-29.80 ± 0.06	6.08 ± 0.06	-185.50 ± 5.42	-144.08 ± 4.88	133.52 ± 6.46	12.6 ± 0.5	1.7 ± 0.1	0.76 ± 0.02	6.8 ± 0.7	0.26 ± 0.02	663.13 ± 65.3	0.89 ± -0.022
236.1068	10.5311	-3.87 ± 0.04	-4.77 ± 0.04	49.11 ± 2.00	-383.69 ± 8.65	18.86 ± 1.48	10.2 ± 0.3	7.3 ± 0.9	0.17 ± 0.04	10.1 ± 0.3	0.14 ± 0.02	4149.48 ± 228.01	0.813 ± -0.038
237.8246	10.1426	-15.19 ± 0.05	-12.74 ± 0.04	-70.00 ± 2.78	-583.25 ± 36.27	34.02 ± 10.0	34.6 ± 18.1	5.8 ± 0.1	0.67 ± 0.12	24.7 ± 15.2	-0.73 ± 0.03	1998.84 ± 520.0	0.348 ± -0.226
237.9600	15.4022	-11.75 ± 0.04	-23.56 ± 0.03	-91.49 ± 2.99	-242.87 ± 6.53	-188.22 ± 0.46	8.3 ± 0.1	4.0 ± 0.1	0.36 ± 0.01	8.3 ± 0.2	-0.06 ± 0.02	2986.96 ± 133.31	1.012 ± -0.006
240.4216	9.6761	1.71 ± 0.03	-11.53 ± 0.02	329.53 ± 14.23	-270.18 ± 13.16	-177.61 ± 9.54	47.9 ± 13.2	3.6 ± 0.4	0.85 ± 0.02	41.5 ± 11.1	0.34 ± 0.03	2968.97 ± 414.55	0.157 ± -0.131
245.4387	8.9954	-25.09 ± 0.04	-23.90 ± 0.03	-2.07 ± 1.29	-316.79 ± 6.09	29.47 ± 1.70	6.8 ± 0.1	1.7 ± 0.1	0.61 ± 0.03	1.5 ± 0.1	-0.28 ± 0.02	143.66 ± 8.72	1.292 ± -0.007
245.5747	6.8844	-0.49 ± 0.04	11.07 ± 0.03	-307.45 ± 11.56	137.48 ± 13.46	5.32 ± 8.20	131.2 ± 56.6	3.2 ± 0.3	0.95 ± 0.02	88.0 ± 33.4	0.51 ± 0.05	1883.49 ± 411.94	-0.285 ± -0.146
245.8356	13.8777	0.98 ± 0.03	-2.39 ± 0.02	-76.79 ± 2.01	-89.61 ± 1.09	-144.63 ± 1.55	6.9 ± 0.1	4.9 ± 0.1	0.17 ± 0.02	6.0 ± 0.1	0.27 ± 0.01	1865.8 ± 97.43	1.055 ± -0.002
246.5144	5.9826	2.42 ± 0.03	1.45 ± 0.02	-5.81 ± 0.40	28.86 ± 0.75	-17.21 ± 0.46	10.1 ± 0.1	6.2 ± 0.0	0.24 ± 0.01	2.2 ± 0.1	0.79 ± 0.0	189.37 ± 9.41	0.873 ± -0.002
246.8588	12.3193	-8.20 ± 0.03	-5.43 ± 0.02	-23.27 ± 1.77	-326.57 ± 13.64	83.74 ± 6.31	6.7 ± 0.7	2.7 ± 0.6	0.43 ± 0.11	5.9 ± 0.6	-0.16 ± 0.01	1489.34 ± 239.52	1.226 ± -0.042
248.4394	7.9229	-7.82 ± 0.05	-20.68 ± 0.04	34.18 ± 1.02	-94.12 ± 1.82	-18.85 ± 0.35	7.5 ± 0.1	3.3 ± 0.1	0.39 ± 0.01	0.6 ± 0.1	0.51 ± 0.01	29.83 ± 1.04	1.155 ± -0.004
248.4959	15.0776	-2.87 ± 0.03	-11.51 ± 0.03	277.89 ± 11.70	-464.93 ± 15.46	-112.07 ± 2.48	56.2 ± 16.2	4.4 ± 0.5	0.85 ± 0.02	47.1 ± 12.6	0.43 ± 0.06	3269.42 ± 292.39	0.077 ± -0.134
250.6963	8.3743	-6.84 ± 0.04	-1.89 ± 0.04	-21.03 ± 0.87	-205.95 ± 7.35	162.04 ± 5.92	7.5 ± 0.3	0.9 ± 0.2	0.80 ± 0.03	7.4 ± 0.4	0.01 ± 0.01	1634.18 ± 172.91	1.176 ± -0.026
251.4082	12.3657	-7.86 ± 0.06	-1.27 ± 0.05	-70.42 ± 2.20	-400.15 ± 11.31	386.54 ± 11.06	201.6 ± 102.4	9.5 ± 0.7	0.90 ± 0.03	155.3 ± 63.8	0.05 ± 0.03	11689.1 ± 785.88	-0.444 ± -0.127
252.1639	15.0648	1.49 ± 0.03	-5.48 ± 0.03	94.04 ± 5.86	-203.86 ± 3.85	-231.61 ± 4.19	15.2 ± 1.1	4.5 ± 0.2	0.54 ± 0.01	14.1 ± 1.0	0.26 ± 0.02	2869.78 ± 110.49	0.721 ± -0.04
253.8582	15.7240	-2.22 ± 0.03	-5.72 ± 0.03	88.00 ± 6.04	-263.85 ± 8.44	-57.11 ± 0.63	7.9 ± 0.6	1.9 ± 0.2	0.62 ± 0.04	5.7 ± 0.5	0.19 ± 0.02	968.04 ± 56.69	1.165 ± -0.031
254.5207	15.4969	14.12 ± 0.02	-9.85 ± 0.03	44.56 ± 0.56	18.18 ± 0.26	-28.06 ± 0.72	11.7 ± 0.1	7.2 ± 0.1	0.24 ± 0.00	0.7 ± 0.1	0.95 ± 0.0	20.13 ± 0.98	0.785 ± -0.001
254.5469	10.9129	-21.80 ± 0.04	-28.68 ± 0.04	260.45 ± 10.30	-383.71 ± 23.23	133.82 ± 4.69	25.0 ± 5.2	2.5 ± 0.3	0.82 ± 0.01	20.4 ± 4.0	-0.28 ± 0.05	1683.88 ± 96.41	0.501 ± -0.11
255.2671	14.9711	-8.81 ± 0.04	-40.50 ± 0.04	208.89 ± 3.70	-210.05 ± 4.37	-32.63 ± 0.96	12.1 ± 0.2	0.7 ± 0.1	0.88 ± 0.02	1.0 ± 0.1	0.16 ± 0.01	51.85 ± 1.75	0.96 ± -0.01
255.5555	10.8612	-5.06 ± 0.04	-15.51 ± 0.04	-108.68 ± 9.34	-418.27 ± 13.88	-213.57 ± 1.85	12.8 ± 1.0	5.5 ± 0.2	0.40 ± 0.04	8.9 ± 0.4	-0.55 ± 0.02	1544.81 ± 66.8	0.781 ± -0.028

Table D.8: Stellar identifications and CaHK$_o$ magnitudes for GRACES targets selected from the *Pristine* survey (Starkenburg et al., 2017b). Also shown are the SDSS *gri* magnitudes, dereddened using E(B-V) values from (Schlegel et al., 1998), and the observer V magnitudes from the SDSS conversions. Total exposure times (number of exposures), observation dates, and final SNR (at 6000 Å) are also provided.

ID	RA (deg)	Dec (deg)	V (mag)	E(B-V)	CaHK$_0$ (mag)	g_0 (mag)	r_0 (mag)	i_0 (mag)	t_{exp} (s, #)	*SNR*	Obs. Dates
2018A:											
P191.8535+12.0508	191.8535	12.0508	15.21	0.03	15.52	15.26	15.05	14.98	5580 (3)	121	6/16, 6/17/2018
P209.0986+09.8244	209.0986	9.8244	15.50	0.03	15.79	15.51	15.32	15.25	6300 (3)	75	6/15/2018
P224.8444+02.3043	224.8444	2.3043	15.21	0.05	15.64	15.27	14.91	14.76	4500 (3)	91	4/24/2018
P237.8589+12.5660	237.8589	12.5660	15.58	0.04	15.84	15.58	15.37	15.30	8100 (4)	95	6/15, 6/16/2018
P244.8986+10.9310	244.8986	10.9310	15.58	0.07	15.76	15.48	15.28	15.22	7200 (4)	86	4/25, 6/15/2018
2018B:											
P021.6938+29.0039	21.6938	29.0039	15.78	0.08	16.00	15.72	15.47	15.36	9600 (4)	61	1/16, 1/17/2019
P031.9938+27.7363	31.9938	27.7363	15.55	0.05	15.77	15.54	15.34	15.27	5400 (3)	74	1/17/2019
P180.3206+02.5788	180.3206	2.5788	15.71	0.02	16.30	15.89	15.46	15.28	9600 (4)	50	1/18, 1/20/2019
P182.5866+09.8940	182.5866	9.8940	15.23	0.02	15.60	15.30	15.07	15.00	5400 (3)	87	1/19/2019
P184.1783+01.0664	184.1783	1.0664	15.81	0.02	16.11	15.85	15.66	15.59	7200 (3)	75	1/16/2019
P184.2997+43.4721	184.2997	43.4721	15.92	0.01	16.70	16.16	15.68	15.49	7200 (3)	46	1/19/2019
P188.0262+00.2055	188.0262	0.2055	15.51	0.02	15.86	15.58	15.34	15.27	7200 (4)	76	1/19, 1/20/2019
P198.0851+08.9428	198.0851	8.9428	15.67	0.03	16.05	15.74	15.47	15.38	7200 (3)	114	1/20/2019
2019A:											
P192.3242+13.3956	192.3242	13.3956	15.61	0.03	16.03	15.69	15.38	15.28	9600 (4)	68	3/27/2019
P194.9935+12.0585	194.9935	12.0585	15.25	0.03	15.81	15.36	15.02	14.91	6000 (3)	53	3/27, 3/28/2019
P207.3454+14.1268	207.3454	14.1268	16.45	0.03	16.77	16.50	16.28	16.18	14400 (6)	54	6/22, 6/28/2019
P207.9290+03.2767	207.9290	3.2767	15.50	0.03	15.79	15.53	15.34	15.27	9600 (4)	81	3/26/2019
P236.9604+11.6155	236.9604	11.6155	15.63	0.06	16.26	15.73	15.24	15.01	8800 (4)	34	3/29/2019

Continued on next page

Table D.8 – *continued from previous page*

ID	RA (deg)	Dec (deg)	V (mag)	E(*B-V*)	$CaHK_0$ (mag)	g_0 (mag)	r_0 (mag)	i_0 (mag)	t_{exp} (s, #)	SNR	Obs. Dates
P246.9682+08.5360	246.9682	8.5360	15.57	0.06	15.77	15.51	15.30	15.24	8800 (4)	44	3/30/2019
P247.2115+06.6348	247.2115	6.6348	15.39	0.06	16.33	15.55	14.92	14.65	6600 (3)	47	3/31/2019
P257.3131+12.8939	257.3131	12.8939	15.31	0.12	15.82	15.24	14.71	14.49	6600 (3)	53	3/29/2019
P258.1048+40.5405	258.1048	40.5405	16.06	0.04	16.34	16.07	15.88	15.82	31200 (13)	53	3/31, 6/22/2019
2019B:											
P008.5638+28.1855	8.5638	28.1855	15.31	0.05	15.65	15.34	15.06	14.97	8100 (3)	105	11/14/2019
P016.2907+28.3957	16.2907	28.3957	14.93	0.06	15.49	15.01	14.59	14.41	5400 (3)	112	11/08/2019
P021.9576+32.4131	21.9576	32.4131	15.75	0.05	16.36	15.89	15.42	15.21	8100 (3)	78	11/11, 11/12/2019
P113.8240+45.1863	113.8240	45.1863	15.43	0.07	15.71	15.39	15.13	15.06	7200 (3)	40	11/09/2019
P116.9657+33.5337	116.9657	33.5337	15.52	0.06	16.16	15.59	15.13	14.91	7200 (3)	55	11/11/2019
P133.0683+28.7219	133.0683	28.7219	15.39	0.03	16.08	15.56	15.11	14.90	7200 (3)	66	01/18/2020
P339.1417+25.5503	339.1417	25.5503	14.77	0.04	15.01	14.76	14.54	14.47	5400 (3)	112	11/08/2019
P339.3203+25.8764	339.3203	25.8764	15.37	0.06	15.78	15.40	15.04	14.90	7200 (3)	66	11/09, 11/11/2019

Table D.9: Stellar parameters from photometry and medium resolution spectroscopy. Temperatures and metallicities are shown from SDSS and CaHK photometry from the *Pristine* survey (Starkenburg et al., 2017a). The stellar parameters (T_{eff}, log g, [Fe/H]$_{MRS}$, and carbonicities are determined from the medium resolution INT spectral survey (Aguado et al., 2019b) are also shown, with the exception of P207.3454+14.1268 which was observed from a medium-resolution ESO 3.6m-EFOSC spectrum (Buzzoni et al., 1984). The C-rich flag denotes whether the star has [C/Fe]> 1.0 based on the MRS FERRE analysis; an asterisk denotes that the SNR was too low for FERRE to determine reliable carbonicities. MRS radial velocities are from Sestito et al. (2020b) and the radial velocities derived from the HRS GRACES spectra are provided for comparison.

ID	PRIS	PRIS	MRS	MRS	MRS	MRS	MRS	MRS	GRACES
	T_{phot}	[Fe/H]$_{phot}$	SNR	T_{eff}	log g	[Fe/H]$_{MRS}$	C-rich?	RV	RV
	(K)	(dex)		(K)	(dex)	(dex)		(km s^{-1})	(km s^{-1})
P008.5638+28.1855	6173	-3.04	34	6169 ± 17	5.00 ± 0.07	-2.84 ± 0.05	N*	-266.4 ± 7.4	-272.3 ± 0.5
P016.2907+28.3957	5583	-3.02	21	5378 ± 24	1.13 ± 0.10	-3.27 ± 0.04	Y	-384.5 ± 10.8	-374.3 ± 0.5
P021.6938+29.0039	6184	-3.31	25	5789 ± 20	1.09 ± 0.09	-3.40 ± 0.04	Y*	-91.0 ± 8.4	-85.4 ± 0.5
P021.9576+32.4131	5408	-3.53	55	5379 ± 147	2.40 ± 0.84	-2.67 ± 0.19	N*	—	-152.6 ± 0.1
P031.9938+27.7363	6465	-3.27	25	6477 ± 35	4.93 ± 0.04	-2.83 ± 0.04	N*	-215.1 ± 11.1	-209.4 ± 0.6
P113.8240+45.1863	6296	-2.71	27	5900 ± 12	5.00 ± 0.00	-3.10 ± 0.03	Y*	-126.9 ± 7.6	-135.9 ± 0.6
P116.9657+33.5337	5423	-2.65	20	5025 ± 15	1.01 ± 0.01	-3.04 ± 0.02	N	-171.5 ± 11.6	-180.9 ± 0.4
P133.0683+28.7219	5445	-3.01	8	5288 ± 29	1.02 ± 0.01	-3.41 ± 0.05	Y*	6.8 ± 20.7	11.9 ± 0.6
P180.3206+02.5788	5563	-3.53	25	5406 ± 107	1.84 ± 1.14	-2.99 ± 0.11	N*	-212.3 ± 9.7	-204.9 ± 0.4
P182.5866+09.8940	6360	-2.87	34	6237 ± 131	5.00 ± 0.44	-3.48 ± 0.11	N*	139.5 ± 8.3	95.3 ± 0.2
P184.1783+01.0664	6483	-2.97	33	6323 ± 48	4.36 ± 0.14	-3.49 ± 0.06	Y*	407.2 ± 10.8	397.6 ± 1.4
P184.2997+43.4721	5419	-2.92	24	5509 ± 26	4.97 ± 0.02	-3.66 ± 0.04	N*	-104.4 ± 12.0	-110.3 ± 0.6
P188.0262+00.2055	6339	-3.00	16	6049 ± 52	4.76 ± 0.18	-3.23 ± 0.07	Y*	114.4 ± 8.7	93.3 ± 0.3

Continued on next page

Table D.9 – *continued from previous page*

ID	PRIS T_{phot} (K)	PRIS $[Fe/H]_{phot}$ (dex)	MRS SNR	MRS T_{eff} (K)	MRS $\log g$ (dex)	MRS $[Fe/H]_{MRS}$ (dex)	MRS C-rich?	MRS RV (km s^{-1})	GRACES RV (km s^{-1})
P191.8535+12.0508	6422	-3.14	33	6270 ± 53	4.75 ± 0.12	-3.88 ± 0.11	N*	18.8 ± 9.9	30.6 ± 0.4
P192.3242+13.3956	6047	-3.10	22	6142 ± 34	4.96 ± 0.03	-2.75 ± 0.04	N*	—	61.1 ± 0.7
P194.9935+12.0585	5936	-2.41	28	5935 ± 39	4.95 ± 0.05	-2.89 ± 0.06	N*	-34.8 ± 7.8	-42.7 ± 3.0
P198.0851+08.9428	6215	-3.00	7	5623 ± 81	1.26 ± 0.25	-3.37 ± 0.11	N*	—	105.1 ± 0.1
P207.3454+14.1268	6304	-3.17	55	6058 ± 280	4.14 ± 0.89	-4.01 ± 0.65	N*	—	-78.8 ± 0.1
P207.9290+03.2767	6516	-2.95	18	6490 ± 46	4.89 ± 0.10	-2.96 ± 0.07	N*	126.4 ± 7.1	109.3 ± 0.3
P209.0986+09.8244	6477	-2.92	32	6395 ± 38	4.85 ± 0.09	-3.14 ± 0.05	N*	-108.8 ± 8.9	-116.7 ± 0.1
P224.8444+02.3043	5791	-3.40	19	5455 ± 39	1.23 ± 0.18	-3.92 ± 0.08	Y*	196.3 ± 24.3	189.2 ± 0.8
P236.9604+11.6155	5320	-3.21	23	5298 ± 29	1.10 ± 0.08	-3.39 ± 0.03	N	-47.8 ± 9.8	-53.5 ± 0.9
P237.8589+12.5660	6443	-3.08	37	6323 ± 34	4.53 ± 0.09	-3.88 ± 0.08	N*	6.5 ± 8.0	-3.8 ± 0.9
P244.8986+10.9310	6492	-2.92	26	6481 ± 28	4.97 ± 0.03	-3.11 ± 0.05	N*	-182.1 ± 11.6	-176.7 ± 0.3
P246.9682+08.5360	6446	-3.13	34	6411 ± 32	4.70 ± 0.07	-3.45 ± 0.05	N*	-55.2 ± 9.8	-58.0 ± 0.5
P247.2115+06.6348	4969	-2.87	9	5026 ± 68	1.96 ± 0.36	-2.99 ± 0.08	N*	—	-200.8 ± 0.4
P257.3131+12.8939	5277	-2.99	19	5281 ± 53	2.50 ± 0.48	-2.80 ± 0.06	N	-129.4 ± 7.0	-129.4 ± 0.4
P258.1048+40.5405	6510	-2.92	79	6283 ± 297	3.95 ± 0.90	-3.76 ± 0.73	N*	—	-169.4 ± 1.5
P339.1417+25.5503	6413	-3.15	27	6048 ± 30	5.00 ± 0.00	-3.42 ± 0.06	Y*	-53.5 ± 7.8	-78.9 ± 0.1
P339.3203+25.8764	5816	-3.24	22	5589 ± 31	4.99 ± 0.01	-3.49 ± 0.07	N*	-141.5 ± 9.0	-157.7 ± 0.3

Table D.10: Stellar Parameters inferred from the Bayesian inference method (see text, Sestito et al. 2020b). Gaia DR2 RA and DEC values, parallaxes, and inferred distances are shown. Also shown are the inferred temperature (T_{eff}) and surface gravity ($\log g$) from the MIST/MESA metal-poor stellar isochrone fitting. Only P031.9938+27.7363 has nearly equal probabilities for the dwarf vs giant solutions, thus both solutions are shown. Short ID's are used to identify stars in a majority of the figures in this paper.

ID	Short ID	RA (deg)	DEC (deg)	ϖ (mas)	D (kpc)	T_{eff} (K)	$\log g$ (dex)
P008.5638+28.1855	P008+28	8.56380914	28.18546890	0.8800 ± 0.0436	1.00 ± 0.03	6297 ± 40	4.53 ± 0.01
P016.2907+28.3957	P016+28	16.29068948	28.39569218	0.0840 ± 0.0365	5.98 ± 0.42	5417 ± 30	2.58 ± 0.07
P021.6938+29.0039	P021+29	21.69382090	29.00384303	0.3643 ± 0.0452	3.33 ± 0.13	6400 ± 133	3.74 ± 0.18
P021.9576+32.4131	P021+32	21.95761802	32.41313561	0.1301 ± 0.0519	12.91 ± 0.70	5254 ± 26	2.19 ± 0.06
P031.9938+27.7363	P031+27	31.99384756	27.73636495	0.4654 ± 0.0553	2.64 ± 0.11	6639 ± 63	3.94 ± 0.05
—	—	—	—	—	1.62 ± 0.11	6632 ± 40	4.33 ± 0.04
P113.8240+45.1863	P113+45	113.82402880	45.18635761	0.7620 ± 0.0506	1.14 ± 0.04	6402 ± 35	4.48 ± 0.04
P116.9657+33.5337	P116+33	116.96571910	33.53367734	0.1269 ± 0.0452	9.52 ± 0.54	5286 ± 27	2.34 ± 0.06
P133.0683+28.7219	P133+28	133.06833770	28.72201402	0.1443 ± 0.0514	10.12 ± 0.60	5297 ± 28	2.29 ± 0.07
P180.3206+02.5788	P180+02	180.32054740	2.57875633	0.1616 ± 0.0608	8.07 ± 0.63	5420 ± 32	2.67 ± 0.08
P182.5866+09.8940	P182+09	182.58641020	9.89386345	0.7807 ± 0.0476	1.16 ± 0.04	6466 ± 33	4.46 ± 0.02
P184.1783+01.0664	P184+01	184.17817900	1.06637007	0.2203 ± 0.0587	3.61 ± 0.26	6425 ± 222	3.75 ± 0.14
P184.2997+43.4721	P184+43	184.29967340	43.47212900	1.4142 ± 0.0506	0.69 ± 0.01	5399 ± 24	4.76 ± 0.00
P188.0262+0.2055	P188+00	188.02636060	0.20531655	0.7269 ± 0.0441	1.27 ± 0.05	6425 ± 34	4.47 ± 0.02
P191.8535+12.0508	P191+12	191.85350510	12.05083083	0.4436 ± 0.0362	2.73 ± 0.06	6687 ± 60	3.82 ± 0.04
P192.3242+13.3956	P192+13	192.32412860	13.39560015	0.2900 ± 0.0435	3.40 ± 0.18	6139 ± 117	3.63 ± 0.11
P194.9935+12.0585	P194+12	194.99344830	12.05849741	0.8620 ± 0.0335	0.93 ± 0.02	6194 ± 27	4.57 ± 0.01

Continued on next page

Table D.10 – *continued from previous page*

ID	Short ID	RA (deg)	DEC (deg)	ϖ (mas)	D (kpc)	T_{eff} (K)	$\log g$ (dex)
P198.0851+08.9428	P198+08	198.08507560	8.94283807	-0.0585 ± 0.0555	3.77 ± 0.08	5997 ± 100	3.50 ± 0.10
P207.3454+14.1268	P207+14	207.34545560	14.12687248	0.5402 ± 0.0814	1.83 ± 0.07	6384 ± 43	4.51 ± 0.02
P207.9290+03.2767	P207+03	207.92897750	3.27665882	0.4321 ± 0.0411	2.51 ± 0.13	6694 ± 38	3.99 ± 0.06
P209.0986+09.8244	P209+09	209.09848420	9.82442890	0.3218 ± 0.0435	2.88 ± 0.23	6536 ± 159	3.83 ± 0.11
P224.8444+02.3043	P224+02	224.84441470	2.30438391	0.0194 ± 0.0397	4.71 ± 0.34	5605 ± 36	2.97 ± 0.08
P236.9604+11.6155	P236+11	236.96043370	11.61547762	0.1705 ± 0.0599	13.80 ± 0.69	5186 ± 25	2.03 ± 0.06
P237.8589+12.5660	P237+12	237.85882390	12.56596867	0.2971 ± 0.0410	3.20 ± 0.14	6654 ± 149	3.80 ± 0.09
P244.8986+10.9310	P244+10	244.89853480	10.93098044	0.3805 ± 0.0588	2.78 ± 0.16	6569 ± 120	3.85 ± 0.08
P246.9682+08.5360	P246+08	246.96813340	8.53604912	0.2747 ± 0.0392	3.17 ± 0.16	6330 ± 200	3.70 ± 0.12
P247.2115+06.6348	P247+06	247.21148800	6.63475417	0.0830 ± 0.0410	19.70 ± 0.55	4892 ± 17	1.48 ± 0.04
P257.3131+12.8939	P257+12	257.31306890	12.89392137	0.0565 ± 0.0370	11.07 ± 0.48	5133 ± 53	2.00 ± 0.05
P258.1048+40.5405	P258+40	258.10480690	40.54046406	0.3890 ± 0.0364	2.04 ± 0.10	6698 ± 43	4.36 ± 0.03
P339.1417+25.5503	P339.1+25.5	339.14175960	25.55028202	0.9142 ± 0.0342	1.02 ± 0.03	6569 ± 24	4.39 ± 0.02
P339.3203+25.8764	P339.3+25.8	339.32028090	25.87639240	0.2368 ± 0.0438	3.86 ± 0.22	5698 ± 43	3.23 ± 0.07

Table D.11: Iron abundances for the sample. Errors represent the total combined systematic error due to the stellar parameters (see Table D.10 and the line-to-line scatter (see Table D.19). The number of lines used to calculate the average [Fe/H] are given in parentheses. ΔFe$_{\mathrm{INLTE}}$ is the averaged NLTE correction for Fe I lines with known NLTE calculations, in the sense that Fe$_{\mathrm{INLTE}}$ = Fe$_{\mathrm{ILTE}}$ + ΔFe$_{\mathrm{INLTE}}$. *Slope* is the slope of the line fit to A(Fe I) vs. excitation potential as a spectroscopic check of the effective temperature (when more than 10 lines are available).

ID	[Fe I/H]	[Fe II/H]	[Fe/H]	FeI−FeII	ΔFe$_{\mathrm{INLTE}}$	*Slope*
P008.5638+28.1855	-2.68 ± 0.06 (22)	-2.89 ± 0.11 (3)	-2.72 ± 0.12 (25)	0.21	0.08 (13)	-0.07
P016.2907+28.3957	-3.12 ± 0.06 (50)	-3.01 ± 0.09 (6)	-3.09 ± 0.07 (56)	-0.11	0.24 (26)	0.03
P021.6938+29.0039	-2.79 ± 0.13 (12)	-2.93 ± 0.19 (4)	-2.80 ± 0.13 (16)	0.14	0.19 (7)	0.03
P021.9576+32.4131	-3.12 ± 0.05 (45)	-3.56 ± 0.08 (3)	-3.24 ± 0.22 (48)	0.44	0.29 (23)	-0.01
P031.9938+27.7363	-2.65 ± 0.09 (13)	-2.95 ± 0.11 (4)	-2.75 ± 0.17 (17)	0.30	0.18 (8)	-0.04
P113.8240+45.1863	-2.14 ± 0.06 (27)	-2.43 ± 0.15 (4)	-2.17 ± 0.16 (31)	0.29	0.07 (13)	0.02
P116.9657+33.5337	-2.82 ± 0.05 (62)	-3.08 ± 0.09 (5)	-2.88 ± 0.14 (67)	0.26	0.21 (32)	-0.03
P133.0683+28.7219	-2.71 ± 0.05 (68)	-3.04 ± 0.08 (6)	-2.79 ± 0.17 (74)	0.33	0.22 (36)	-0.05
P180.3206+02.5788	-3.07 ± 0.07 (37)	-3.36 ± 0.14 (3)	-3.10 ± 0.16 (40)	0.29	0.23 (22)	-0.03
P182.5866+09.8940	-2.81 ± 0.07 (7)	-3.14 ± 0.17 (3)	-2.85 ± 0.18 (10)	0.33	0.12 (5)	-0.02
P184.1783+01.0664	-3.37 ± 0.25 (3)	-3.72 ± 0.46 (1)	-3.43 ± 0.28 (4)	0.35	0.28 (1)	2.44
P184.2997+43.4721	-3.16 ± 0.08 (14)	<-3.30	-3.23 ± 0.11 (17)	>0.14	0.02 (6)	-0.10
P188.0262+00.2055	-2.81 ± 0.06 (10)	-3.11 ± 0.11 (3)	-2.88 ± 0.16 (13)	0.30	0.11 (4)	-0.03
P191.8535+12.0508	-3.22 ± 0.11 (7)	-3.75 ± 0.12 (3)	-3.42 ± 0.28 (10)	0.53	0.28 (2)	0.04
P192.3242+13.3956	-2.38 ± 0.10 (42)	-2.66 ± 0.12 (4)	-2.42 ± 0.16 (46)	0.28	0.15 (25)	-0.04
P194.9935+12.0585	-2.81 ± 0.22 (8)	-2.97 ± 0.28 (2)	-2.82 ± 0.19 (10)	0.16	0.08 (5)	-0.06
P198.0851+08.9428	-2.54 ± 0.11 (24)	-2.02 ± 0.15 (6)	-2.49 ± 0.27 (30)	-0.52	0.15 (14)	0.01

Continued on next page

Table D.11 – *continued from previous page*

ID	[Fe I/H]	[Fe II/H]	[Fe/H]	FeI−FeII	ΔFe$_{INLTE}$	*Slope*
P207.3454+14.1268	-3.20 ± 0.10 (2)	-3.21 ± 0.23 (2)	-3.20 ± 0.09 (4)	0.01	0.13 (1)	-0.89
P207.9290+03.2767	-2.81 ± 0.06 (8)	-3.14 ± 0.11 (3)	-2.88 ± 0.17 (11)	0.33	0.20 (4)	-0.07
P209.0986+09.8244	-2.93 ± 0.16 (8)	-3.24 ± 0.12 (3)	-2.99 ± 0.18 (11)	0.31	0.22 (4)	0.00
P224.8444+02.3043	-3.65 ± 0.10 (7)	-3.47 ± 0.12 (3)	-3.62 ± 0.12 (10)	-0.18	0.30 (4)	0.04
P236.9604+11.6155	-3.50 ± 0.10 (11)	-3.83 ± 0.08 (3)	-3.70 ± 0.18 (14)	0.33	0.38 (6)	0.12
P237.8589+12.5660	<-3.73	<-4.03	<-3.88	<0.30	0.36 (1)	0.00
P244.8986+10.9310	-3.09 ± 0.13 (6)	-3.42 ± 0.12 (3)	-3.15 ± 0.19 (9)	0.33	0.23 (3)	-0.04
P246.9682+08.5360	<-3.40	<-3.80	<-3.60	<0.40	0.27 (9)	0.29
P247.2115+06.6348	-3.20 ± 0.05 (45)	-3.43 ± 0.10 (4)	-3.25 ± 0.12 (49)	0.23	0.25 (22)	-0.08
P257.3131+12.8939	-2.85 ± 0.07 (63)	-3.21 ± 0.11 (3)	-2.91 ± 0.19 (66)	0.36	0.23 (33)	-0.04
P258.1048+40.5405	<-3.25	<-3.80	<-3.52	<0.55	0.20 (8)	0.25
P339.1417+25.5503	-2.66 ± 0.06 (19)	-2.82 ± 0.09 (4)	-2.69 ± 0.09 (23)	0.16	0.12 (13)	-0.02
P339.3203+25.8764	-3.23 ± 0.07 (23)	-3.36 ± 0.09 (3)	-3.26 ± 0.09 (26)	0.13	0.22 (12)	-0.00
HD 122563	-2.90 ± 0.09 (120)	-2.63 ± 0.07 (9)	-2.82 ± 0.15 (129)	-0.27	0.11 (48)	-0.04

Table D.12: LTE abundances for the α-elements. Errors represent the total combined systematic error due to the stellar parameters and line-to-line scatter (Table D.16 and D.17). The number of lines used is given in parentheses.

ID	[O I/H]	[Mg I/H]	[Ca I/H]	[Ca II/H]	[Ti I/H]	[Ti II/H]
P008.5638+28.1855	<-2.02	-2.32 ± 0.06 (4)	-2.64 ± 0.07 (3)	-2.03 ± 0.09 (3)	<-2.22	<-2.17
P016.2907+28.3957	<-1.79	-2.70 ± 0.14 (4)	-2.76 ± 0.08 (10)	-1.97 ± 0.14 (3)	-2.51 ± 0.08 (4)	-2.35 ± 0.07 (6)
P021.6938+29.0039	<-2.35	-2.98 ± 0.14 (2)	-2.66 ± 0.11 (4)	-2.13 ± 0.17 (3)	<-1.70	<-2.45
P021.9576+32.4131	<-2.34	-2.56 ± 0.12 (3)	-2.78 ± 0.08 (7)	-2.75 ± 0.11 (3)	<-3.04	-2.99 ± 0.19 (1)
P031.9938+27.7363	<-2.35	-2.47 ± 0.11 (3)	-2.43 ± 0.09 (4)	-1.94 ± 0.06 (3)	<-1.85	<-2.15
P113.8240+45.1863	<-1.87	-2.17 ± 0.25 (2)	-2.18 ± 0.12 (5)	-1.87 ± 0.16 (1)	<-1.52	<-1.67
P116.9657+33.5337	-2.08 ± 0.25 (1)	-2.39 ± 0.08 (3)	-2.58 ± 0.08 (8)	-2.51 ± 0.20 (3)	-2.50 ± 0.08 (3)	-2.55 ± 0.11 (2)
P133.0683+28.7219	<-1.94	-2.91 ± 0.35 (2)	-2.57 ± 0.07 (10)	-2.24 ± 0.15 (3)	-2.49 ± 0.16 (2)	-2.44 ± 0.12 (1)
P180.3206+02.5788	<-1.95	-2.75 ± 0.13 (3)	-2.81 ± 0.08 (6)	-2.47 ± 0.14 (3)	-2.53 ± 0.18 (2)	<-1.70
P182.5866+09.8940	<-1.95	-3.05 ± 0.09 (2)	<-2.05	-2.27 ± 0.09 (3)	<-2.00	<-1.95
P184.1783+01.0664	-1.88 ± 0.19 (2)	-3.10 ± 0.15 (2)	<-3.03	-2.29 ± 0.11 (2)	<-2.08	<-2.23
P184.2997+43.4721	<-1.03	-2.60 ± 0.21 (2)	<-3.48	-3.16 ± 0.11 (3)	<-2.38	<-1.88
P188.0262+00.2055	<-1.98	-2.61 ± 0.09 (3)	-2.73 ± 0.21 (2)	-2.11 ± 0.10 (3)	<-1.93	<-2.03
P191.8535+12.0508	<-2.62	-3.30 ± 0.08 (2)	<-3.37	-2.81 ± 0.40 (3)	<-2.02	<-2.42
P192.3242+13.3956	<-2.27	-2.43 ± 0.12 (3)	-2.41 ± 0.11 (5)	-1.82 ± 0.22 (2)	<-2.07	-2.12 ± 0.20 (1)
P194.9935+12.0585	<-1.82	-2.78 ± 0.48 (2)	-2.75 ± 0.20 (2)	-2.72 ± 1.07 (2)	<-2.27	<-1.92
P198.0851+08.9428	-2.03 ± 0.18 (2)	-2.09 ± 0.15 (2)	-2.68 ± 0.12 (4)	-1.46 ± 0.17 (3)	<-2.24	-2.02 ± 0.14 (2)
P207.3454+14.1268	-1.64 ± 0.19 (1)	-2.18 ± 0.07 (3)	<-3.65	-2.32 ± 0.20 (2)	<-1.80	<-1.90
P207.9290+03.2767	-2.48 ± 0.25 (1)	-2.88 ± 0.10 (2)	-2.44 ± 0.15 (3)	-2.00 ± 0.08 (3)	<-1.88	<-2.03
P209.0986+09.8244	<-2.34	-2.60 ± 0.21 (4)	-2.74 ± 0.21 (2)	-2.30 ± 0.24 (3)	<-1.89	<-2.19
P224.8444+02.3043	-1.97 ± 0.65 (1)	-3.47 ± 0.13 (3)	-3.37 ± 0.12 (1)	-2.28 ± 0.19 (3)	<-2.67	<-2.87
P236.9604+11.6155	<-1.88	-3.38 ± 0.29 (2)	-3.15 ± 0.19 (2)	-2.98 ± 0.13 (3)	<-2.65	<-2.85
P237.8589+12.5660	<-2.48	-3.80 ± 0.16 (2)	<-2.98	-3.47 ± 0.36 (3)	<-2.48	<-2.48
P244.8986+10.9310	<-2.40	-3.04 ± 0.35 (4)	-2.70 ± 0.15 (3)	-2.26 ± 0.23 (3)	<-2.00	<-2.25

Continued on next page

Table D.12 – *continued from previous page*

ID	[O I/H]	[Mg I/H]	[Ca I/H]	[Ca II/H]	[Ti I/H]	[Ti II/H]
P246.9682+08.5360	<-2.04	-3.42 ± 0.17 (2)	<-2.80	-2.50 ± 0.11 (3)	<-2.20	<-2.25
P247.2115+06.6348	<-1.99	-2.80 ± 0.12 (3)	-3.01 ± 0.09 (5)	-2.81 ± 0.33 (2)	-2.82 ± 0.11 (4)	-2.73 ± 0.18 (4)
P257.3131+12.8939	<-2.01	-2.33 ± 0.09 (4)	-2.56 ± 0.08 (8)	-2.41 ± 0.11 (3)	-2.49 ± 0.11 (3)	-2.54 ± 0.09 (5)
P258.1048+40.5405	<-2.31	-3.21 ± 0.13 (2)	<-2.91	-2.15 ± 0.20 (2)	<-1.96	<-1.96
P339.1417+25.5503	-2.29 ± 0.15 (2)	-2.66 ± 0.11 (2)	-2.47 ± 0.10 (5)	-1.84 ± 0.09 (3)	-2.04 ± 0.18 (1)	<-2.34
P339.3203+25.8764	<-1.96	-2.91 ± 0.13 (2)	-3.06 ± 0.06 (4)	-2.64 ± 0.11 (3)	<-2.68	<-2.71
HD 122563	—	-2.40 ± 0.06 (4)	-2.66 ± 0.07 (18)	—	-2.74 ± 0.10 (16)	-2.37 ± 0.07 (10)

Table D.13: LTE abundances for light elements and Fe-peak elements. Errors represent the total combined systematic error due to the stellar parameters and line-to-line scatter (Table D.18 and D.19). The number of lines used is given in parentheses. P016.2907+28.3957 is the only star in the sample to have a detectable Sc abundance. The systematic errors for P016.2907+28.3957 are given in Table D.21.

ID	[Na I/H]	[K I/H]	[Sc II/H]	[Cr I/H]	[Mn I/H]	[Ni I/H]	[Cu I/H]	[Zn I/H]
P008.5638+28.1855	-2.97 ± 0.08 (1)	<-2.12	<-1.12	-2.87 ± 0.11 (1)	<-2.32	<-2.62	<-1.62	<-1.77
P016.2907+28.3957	-3.15 ± 0.07 (2)	<-2.54	-2.24 ± 0.36 (1)	-3.17 ± 0.13 (4)	<-2.89	-3.09 ± 0.09 (1)	<-2.54	<-2.49
P021.6938+29.0039	<-3.30	<-2.30	<-0.80	-2.80 ± 0.21 (1)	<-1.85	-2.60 ± 0.25 (1)	<-1.40	<-1.30
P021.9576+32.4131	-3.30 ± 0.08 (2)	-2.49 ± 0.16 (1)	<-2.04	-3.30 ± 0.08 (3)	<-2.84	-3.19 ± 0.12 (1)	<-2.44	<-2.24
P031.9938+27.7363	-2.90 ± 0.08 (1)	<-1.85	<-0.75	-2.80 ± 0.19 (1)	<-1.90	<-2.50	<-1.35	<-1.55
P113.8240+45.1863	-2.01 ± 0.20 (2)	<-1.87	<-0.37	-2.42 ± 0.14 (2)	<-1.72	<-2.37	<-1.12	<-1.12
P116.9657+33.5337	-3.03 ± 0.14 (2)	-2.43 ± 0.15 (1)	<-1.88	-3.08 ± 0.09 (3)	<-2.63	-2.93 ± 0.13 (1)	<-2.28	<-2.18
P133.0683+28.7219	-2.94 ± 0.10 (2)	-2.39 ± 0.16 (1)	<-1.89	-3.09 ± 0.08 (3)	<-2.69	-2.81 ± 0.27 (2)	<-2.29	<-2.09
P180.3206+02.5788	-3.35 ± 0.10 (2)	-2.55 ± 0.20 (1)	<-1.10	-3.30 ± 0.14 (1)	<-2.50	<-3.15	<-2.05	<-1.90
P182.5866+09.8940	-3.40 ± 0.13 (2)	<-2.10	<-0.35	<-3.00	<-1.95	<-2.65	<-1.45	<-1.55
P184.1783+01.0664	-2.43 ± 0.62 (2)	<-1.93	<-1.03	<-3.18	<-2.08	<-2.68	<-1.53	<-1.63
P184.2997+43.4721	-3.40 ± 0.09 (2)	<-2.38	<-0.23	-3.58 ± 0.19 (1)	<-2.28	<-3.03	<-2.03	<-1.53
P188.0262+00.2055	-3.38 ± 0.09 (2)	<-2.13	<-0.88	<-3.03	<-2.03	<-2.63	<-1.48	<-1.48
P191.8535+12.0508	<-3.47	<-2.12	<-1.22	<-3.12	<-2.12	<-2.77	<-1.47	<-1.67
P192.3242+13.3956	-3.12 ± 0.18 (2)	<-2.07	<-1.12	-2.59 ± 0.18 (2)	<-2.12	-2.32 ± 0.12 (1)	<-1.57	<-1.62
P194.9935+12.0585	<-3.47	<-1.97	<-0.42	<-3.07	<-2.07	<-2.72	<-1.52	<-1.37
P198.0851+08.9428	-2.79 ± 0.11 (2)	<-2.24	<-1.04	-2.75 ± 0.15 (2)	<-2.19	-2.69 ± 0.19 (1)	<-1.64	<-1.79
P207.3454+14.1268	-2.80 ± 0.11 (2)	<-1.80	<-0.70	<-3.00	<-1.78	<-2.45	<-1.32	<-1.40
P207.9290+03.2767	-3.23 ± 0.12 (2)	<-1.83	<-0.88	-2.73 ± 0.13 (2)	<-1.83	<-2.43	<-1.23	<-1.46
P209.0986+09.8244	-3.08 ± 0.13 (2)	<-2.04	<-1.09	<-3.09	<-1.99	-2.64 ± 0.25 (1)	<-1.39	<-1.54
P224.8444+02.3043	-3.43 ± 0.09 (2)	<-2.22	<-1.52	-3.72 ± 0.17 (1)	<-2.52	<-3.32	<-2.17	<-1.97
P236.9604+11.6155	<-3.00	<-2.70	<-1.20	<-3.80	<-2.45	<-3.35	<-2.16	<-2.20
P237.8589+12.5660	-3.88 ± 0.21 (1)	<-1.98	<-1.08	<-2.98	<-2.03	<-2.88	<-1.33	<-1.53
P244.8986+10.9310	-3.25 ± 0.13 (2)	<-1.95	<-1.15	<-3.00	<-2.00	<-2.55	<-1.40	<-1.60
P246.9682+08.5360	<-3.40	-1.60 ± 0.18 (1)	<-0.60	<-2.80	<-1.75	<-2.60	<-1.16	<-1.30
P247.2115+06.6348	-3.40 ± 0.11 (2)	<-2.80	<-1.85	-3.42 ± 0.13 (4)	<-2.90	-3.30 ± 0.12 (1)	<-2.54	<-2.13
P257.3131+12.8939	-3.04 ± 0.28 (2)	<-2.36	<-1.91	-2.96 ± 0.17 (3)	<-2.76	-2.82 ± 0.25 (2)	<-2.46	<-2.11
P258.1048+40.5405	-3.61 ± 0.16 (1)	<-1.71	<-0.26	<-2.96	<-1.56	<-2.36	<-1.01	<-1.06
P339.1417+25.5503	<-3.14	<-2.09	<-1.09	-2.77 ± 0.11 (2)	<-2.09	<-2.74	<-1.54	<-1.69

Continued on next page

Table D.13 – *continued from previous page*

ID	[Na I/H]	[K I/H]	[Sc II/H]	[Cr I/H]	[Mn I/H]	[Ni I/H]	[Cu I/H]	[Zn I/H]
P339.3203+25.8764	-3.25 ± 0.07 (2)	<-2.66	<-1.66	-3.41 ± 0.17 (1)	<-2.56	<-3.36	<-2.21	<-2.06
HD 122563	-2.84 ± 0.10 (2)	—	—	-3.10 ± 0.11 (8)	-3.20 ± 0.09 (3)	-2.72 ± 0.07 (18)	<-4.02	-2.52 ± 0.07 (1)

Table D.14: LTE abundances for neutron-capture elements. Errors represent the total combined systematic error due to the stellar parameters and line-to-line scatter (Table D.20). P016.2907+28.3957 is the only star in the sample to have detectable La and Nd abundances. The systematic errors for P016.2907+28.3957 are given in Table D.21.

ID	[Y II/H]	[Zr II/H]	[Ba II/H]	[La II/H]	[Nd II/H]	[Eu II/H]
P008.5638+28.1855	<-1.57	<-0.12	<-2.67	<-0.42	<-0.22	<-0.17
P016.2907+28.3957	<-2.69	<-1.19	-1.09 ± 0.30 (2)	-1.46 ± 0.22 (1)	-1.36 ± 0.24 (1)	<-1.04
P021.6938+29.0039	<-1.40	<0.25	<-2.75	<-0.30	<0.30	<-0.30
P021.9576+32.4131	<-2.59	<-0.99	<-3.84	<-1.64	<-1.54	<-1.24
P031.9938+27.7363	<-1.45	<0.15	<-2.50	<-0.25	<0.00	<-0.10
P113.8240+45.1863	<-0.97	<0.73	-1.42 ± 0.13 (2)	<0.18	<0.38	<0.18
P116.9657+33.5337	<-2.43	<-0.83	<-3.73	<-1.48	<-1.28	<-1.18
P133.0683+28.7219	<-2.59	<-0.94	-2.84 ± 0.14 (2)	<-1.49	<-1.29	<-1.14
P180.3206+02.5788	<-2.20	<-0.50	<-3.55	<-1.20	<-1.00	<-0.95
P182.5866+09.8940	<-1.35	<0.15	<-2.50	<-0.25	<0.05	<-0.05
P184.1783+01.0664	<-1.73	<-0.03	-1.98 ± 0.17 (2)	<-0.53	<-0.23	<-0.13
P184.2997+43.4721	<-1.33	<0.37	<-2.83	<-0.43	<-0.13	<-0.18
P188.0262+00.2055	<-1.28	<0.12	<-2.43	<-0.18	<0.02	<0.02
P191.8535+12.0508	<-1.67	<-0.07	<-2.77	<-0.52	<-0.22	<-0.32
P192.3242+13.3956	<-1.62	<-0.12	<-2.87	<-0.57	<-0.22	<-0.42
P194.9935+12.0585	<-1.17	<0.53	<-2.62	<-0.12	<0.13	<-0.12
P198.0851+08.9428	<-2.49	<-0.24	<-2.84	<-0.49	<-0.44	<-0.39
P207.3454+14.1268	<-1.18	<0.42	<-2.40	<-0.21	<0.22	<0.10
P207.9290+03.2767	<-1.38	<0.12	<-2.58	<-0.19	<0.09	<-0.13

Continued on next page

Table D.14 – *continued from previous page*

ID	[Y II/H]	[Zr II/H]	[Ba II/H]	[La II/H]	[Nd II/H]	[Eu II/H]
P209.0986+09.8244	<-1.49	<0.16	<-2.64	<-0.34	<-0.09	<-0.19
P224.8444+02.3043	<-2.17	<-0.42	<-3.32	<-1.17	<-0.92	<-0.92
P236.9604+11.6155	<-2.24	<-0.70	<-3.70	<-1.42	<-1.08	<-1.20
P237.8589+12.5660	<-1.53	<0.12	<-2.58	<-0.48	<-0.13	<-0.18
P244.8986+10.9310	<-1.55	<-0.05	<-2.70	<-0.55	<-0.15	<-0.15
P246.9682+08.5360	<-1.30	<0.20	<-2.55	<-0.25	<0.18	<-0.10
P247.2115+06.6348	<-2.73	<-1.15	-3.90 ± 0.13 (2)	<-1.90	<-1.75	<-1.70
P257.3131+12.8939	<-2.56	<-1.01	-3.28 ± 0.13 (2)	<-1.61	<-1.41	<-1.36
P258.1048+40.5405	<-0.96	<0.64	<-2.32	<0.14	<0.47	<0.13
P339.1417+25.5503	<-1.49	<-0.09	<-2.59	<-0.29	<-0.09	<-0.14
P339.3203+25.8764	<-2.11	<-0.66	<-3.46	<-1.11	<-0.91	<-0.81
HD 122563	-2.99 ± 0.11 (4)	—	-3.72 ± 0.14 (2)	<-2.82	<-2.87	<-2.17

Table D.15 LTE lithium abundances from the Li doublet at 6707 Å. The σ represents the line-to-line scatter for a given element, added in quadrature with errors imposed by the continuum placement for each line used. ΔT and Δg are the systematic errors in the stellar parameters T_{eff} and $\log g$ from Table D.10, while ΔFe is due to the uncertainty in [Fe/H] from Table D.11. Δ is the total combined error of these added in quadrature.

ID	A(Li)	σ	ΔT	Δg	ΔFe	Δ
P008.5638+28.1855	2.06	0.13	-0.05	0.00	-0.01	0.14
P016.2907+28.3957	<0.81					
P021.6938+29.0039	2.00	0.14	-0.10	0.00	-0.04	0.18
P021.9576+32.4131	0.76	0.32	-0.05	0.00	-0.02	0.32
P031.9938+27.7363	2.10	0.18	-0.05	0.00	-0.01	0.19
P113.8240+45.1863	2.08	0.20	-0.05	0.00	0.00	0.20
P116.9657+33.5337	0.92	0.22	0.00	0.00	0.01	0.22
P133.0683+28.7219	<0.71					
P180.3206+02.5788	1.00	0.25	-0.05	0.00	-0.05	0.26
P182.5866+09.8940	1.90	0.16	-0.00	0.00	-0.02	0.16
P184.1783+01.0664	<1.72					
P184.2997+43.4721	<1.22					
P188.0262+00.2055	1.82	0.20	-0.05	0.00	-0.01	0.20
P191.8535+12.0508	2.03	0.18	0.00	0.00	0.02	0.18
P192.3242+13.3956	2.13	0.14	-0.05	0.00	0.00	0.15
P194.9935+12.0585	1.73	0.18	0.00	0.00	0.00	0.18
P198.0851+08.9428	<1.51					
P207.3454+14.1268	<1.80					
P207.9290+03.2767	2.09	0.15	0.00	0.00	0.03	0.15
P209.0986+09.8244	1.96	0.20	-0.05	0.05	0.04	0.21
P224.8444+02.3043	<1.13					
P236.9604+11.6155	<1.04					
P237.8589+12.5660	1.92	0.20	-0.05	0.05	0.05	0.22
P244.8986+10.9310	2.05	0.16	-0.05	0.00	0.03	0.17
P246.9682+08.5360	2.08	0.22	0.05	0.20	0.20	0.36
P247.2115+06.6348	<0.58					
P257.3131+12.8939	<0.79					
P258.1048+40.5405	2.04	0.20	-0.05	0.00	0.00	0.20
P339.1417+25.5503	2.01	0.14	0.00	0.00	0.01	0.14
P339.3203+25.8764	1.49	0.14	0.00	0.00	0.02	0.14
HD 122563	-0.97	0.30	-0.05	0.00	0.01	0.30

Table D.16: Systematic errors for the α-elements (pt. 1). The σ represents the line-to-line scatter for a given element, added in quadrature with errors imposed by the continuum placement for each line used. When the number of lines $N_X > 5$ for species X, then σ is reduced by $1/\sqrt{N_X}$. ΔT and Δg are the systematic errors in the stellar parameters T_{eff} and $\log g$ from Table D.10, while ΔFe is due to the uncertainty in [Fe/H] from Table D.11. Δ is the total combined error of these added in quadrature.

ID	[O I/H]					[Mg I/H]					[Ca I/H]				
	σ	ΔT	Δg	ΔFe	Δ	σ	ΔT	Δg	ΔFe	Δ	σ	ΔT	Δg	ΔFe	Δ
P008.5638+28.1855						0.05	-0.03	0.00	-0.01	0.06	0.07	0.00	0.00	-0.01	0.07
P016.2907+28.3957						0.13	-0.04	-0.01	-0.02	0.14	0.08	-0.00	0.00	0.00	0.08
P021.6938+29.0039						0.10	-0.08	-0.03	-0.02	0.14	0.09	-0.05	0.00	0.01	0.11
P021.9576+32.4131						0.11	-0.02	-0.02	-0.06	0.12	0.06	-0.01	0.00	0.04	0.08
P031.9938+27.7363						0.11	-0.01	0.00	0.01	0.11	0.09	0.02	0.00	0.01	0.09
P113.8240+45.1863						0.25	0.00	0.01	-0.01	0.25	0.12	-0.02	0.00	-0.01	0.12
P116.9657+33.5337	0.24	0.03	-0.02	0.04	0.25	0.08	-0.00	0.00	0.01	0.08	0.08	-0.01	-0.01	0.00	0.08
P133.0683+28.7219						0.33	-0.05	-0.08	-0.02	0.35	0.07	-0.00	0.00	0.01	0.07
P180.3206+02.5788						0.13	-0.01	-0.01	-0.01	0.13	0.08	-0.01	0.00	0.00	0.08
P182.5866+09.8940						0.08	-0.02	0.00	-0.02	0.09					
P184.1783+01.0664	0.13	0.12	-0.05	0.00	0.19	0.11	-0.10	0.00	0.01	0.15					
P184.2997+43.4721						0.21	0.00	0.00	-0.00	0.21					
P188.0262+00.2055						0.09	-0.02	-0.02	-0.01	0.09	0.21	-0.03	0.00	-0.01	0.21
P191.8535+12.0508						0.06	-0.05	0.00	-0.03	0.08					
P192.3242+13.3956						0.10	-0.06	-0.02	-0.01	0.12	0.07	-0.08	-0.02	-0.02	0.11
P194.9935+12.0585						0.19	0.25	0.25	0.25	0.48	0.16	0.08	0.07	0.07	0.20
P198.0851+08.9428	0.15	0.06	-0.07	0.00	0.18	0.13	-0.08	0.00	-0.01	0.15	0.10	-0.05	0.03	0.02	0.12

Continued on next page

Table D.16 – *continued from previous page*

ID	[O I/H]					[Mg I/H]					[Ca I/H]				
	σ	ΔT	Δg	ΔFe	Δ	σ	ΔT	Δg	ΔFe	Δ	σ	ΔT	Δg	ΔFe	Δ
P207.3454+14.1268	0.18	0.04	-0.01	0.01	0.19	0.07	-0.02	-0.01	0.00	0.07					
P207.9290+03.2767	0.20	0.04	0.04	-0.13	0.25	0.10	-0.02	0.00	-0.02	0.10	0.14	-0.04	0.00	-0.02	0.15
P209.0986+09.8244						0.14	-0.13	-0.08	-0.05	0.21	0.19	-0.10	0.00	-0.01	0.21
P224.8444+02.3043	0.24	-0.31	-0.31	-0.41	0.65	0.11	-0.04	0.03	-0.02	0.13	0.11	0.00	0.05	0.00	0.12
P236.9604+11.6155						0.29	0.00	-0.02	-0.01	0.29	0.19	0.00	0.00	0.01	0.19
P237.8589+12.5660						0.11	-0.10	-0.02	-0.05	0.16					
P244.8986+10.9310						0.17	-0.21	-0.15	-0.16	0.35	0.13	-0.05	0.00	0.04	0.15
P246.9682+08.5360						0.10	-0.13	0.00	0.00	0.17					
P247.2115+06.6348						0.11	0.00	0.00	0.03	0.12	0.08	-0.03	0.00	-0.00	0.09
P257.3131+12.8939						0.09	-0.02	-0.01	0.00	0.09	0.07	-0.03	0.00	0.01	0.08
P258.1048+40.5405						0.13	0.00	0.00	0.00	0.13					
P339.1417+25.5503	0.15	0.03	-0.02	-0.01	0.15	0.11	0.00	0.00	0.01	0.11	0.09	-0.04	0.00	-0.03	0.10
P339.3203+25.8764						0.12	0.00	-0.03	-0.01	0.13	0.06	0.00	0.00	-0.00	0.06
HD 122563						0.06	-0.02	0.00	0.01	0.06	0.05	-0.05	0.00	0.01	0.07

Table D.17: Systematic errors for the α-elements (pt. 2). See the caption for Table D.16.

ID	[Ca II/H]					[Ti I/H]					[Ti II/H]				
	σ	ΔT	Δg	ΔFe	Δ	σ	ΔT	Δg	ΔFe	Δ	σ	ΔT	Δg	ΔFe	Δ
P008.5638+28.1855	0.09	-0.03	0.00	0.01	0.09										
P016.2907+28.3957	0.13	-0.04	-0.04	-0.01	0.14	0.07	-0.02	0.00	0.01	0.08	0.06	-0.01	-0.04	-0.01	0.07
P021.6938+29.0039	0.15	-0.03	-0.07	0.01	0.17										
P021.9576+32.4131	0.08	-0.04	-0.05	-0.02	0.11						0.18	0.00	-0.05	-0.02	0.19
P031.9938+27.7363	0.06	0.00	-0.03	-0.01	0.06										
P113.8240+45.1863	0.16	-0.05	0.00	0.00	0.16										
P116.9657+33.5337	0.20	-0.00	-0.03	-0.01	0.20	0.07	-0.02	0.00	0.01	0.08	0.11	0.00	-0.01	-0.00	0.11
P133.0683+28.7219	0.14	0.00	-0.01	0.02	0.15	0.15	-0.05	0.00	-0.02	0.16	0.11	0.00	0.00	0.03	0.12
P180.3206+02.5788	0.13	-0.02	-0.04	0.01	0.14	0.18	-0.03	0.00	0.00	0.18					
P182.5866+09.8940	0.09	-0.01	-0.01	-0.02	0.09										
P184.1783+01.0664	0.10	-0.00	-0.05	0.01	0.11										
P184.2997+43.4721	0.11	-0.02	0.00	-0.02	0.11										
P188.0262+00.2055	0.10	0.00	0.00	-0.01	0.10										
P191.8535+12.0508	0.20	-0.19	-0.22	-0.20	0.40										
P192.3242+13.3956	0.22	0.01	0.01	0.00	0.22						0.18	-0.05	-0.05	0.00	0.20
P194.9935+12.0585	0.37	0.60	0.60	0.54	1.07										
P198.0851+08.9428	0.10	-0.08	-0.12	-0.01	0.17						0.12	-0.02	-0.05	-0.02	0.14
P207.3454+14.1268	0.20	0.00	0.00	0.02	0.20										
P207.9290+03.2767	0.07	0.00	-0.01	0.02	0.08										

Continued on next page

Table D.17 – *continued from previous page*

ID	[Ca II/H]					[Ti I/H]					[Ti II/H]				
	σ	ΔT	Δg	ΔFe	Δ	σ	ΔT	Δg	ΔFe	Δ	σ	ΔT	Δg	ΔFe	Δ
P209.0986+09.8244	0.12	-0.13	-0.14	-0.10	0.24										
P224.8444+02.3043	0.08	-0.12	-0.09	-0.08	0.19										
P236.9604+11.6155	0.12	-0.01	-0.02	0.01	0.13										
P237.8589+12.5660	0.15	-0.18	-0.22	-0.15	0.36										
P244.8986+10.9310	0.19	-0.07	-0.08	-0.05	0.23										
P246.9682+08.5360	0.10	-0.04	-0.01	0.00	0.11										
P247.2115+06.6348	0.33	0.00	-0.04	-0.02	0.33	0.10	0.01	0.01	0.04	0.11	0.18	0.00	-0.01	0.02	0.18
P257.3131+12.8939	0.09	-0.03	-0.05	-0.03	0.11	0.08	-0.07	0.00	-0.02	0.11	0.08	-0.00	-0.00	0.02	0.09
P258.1048+40.5405	0.20	0.00	0.00	0.00	0.20										
P339.1417+25.5503	0.09	0.00	0.00	0.01	0.09	0.18	0.00	0.00	0.01	0.18					
P339.3203+25.8764	0.08	-0.05	-0.05	-0.03	0.11										
HD 122563						0.06	-0.08	0.00	0.00	0.10	0.06	-0.02	-0.02	-0.01	0.07

Table D.18: Systematic errors for [Na I/H] and [K I/H]. See the caption for Table D.16.

ID	[Na I/H]					[K I/H]				
	σ	ΔT	Δg	ΔFe	Δ	σ	ΔT	Δg	ΔFe	Δ
P008.5638+28.1855	0.06	-0.05	0.00	-0.01	0.08					
P016.2907+28.3957	0.07	-0.01	0.00	-0.02	0.07					
P021.6938+29.0039										
P021.9576+32.4131	0.07	-0.05	0.00	-0.02	0.08	0.15	-0.05	0.00	-0.02	0.16
P031.9938+27.7363	0.06	-0.05	0.00	-0.01	0.08					
P113.8240+45.1863	0.20	-0.03	0.00	0.00	0.20					
P116.9657+33.5337	0.12	-0.05	-0.00	-0.04	0.14	0.15	0.00	0.00	0.01	0.15
P133.0683+28.7219	0.09	-0.02	-0.02	0.01	0.10	0.15	0.00	0.00	-0.02	0.16
P180.3206+02.5788	0.10	-0.03	0.00	0.00	0.10	0.20	-0.05	0.00	0.00	0.20
P182.5866+09.8940	0.13	-0.02	0.00	-0.02	0.13					
P184.1783+01.0664	0.09	-0.42	-0.32	-0.31	0.62					
P184.2997+43.4721	0.08	0.00	0.00	0.01	0.09					
P188.0262+00.2055	0.09	0.00	0.00	-0.01	0.09					
P191.8535+12.0508										
P192.3242+13.3956	0.16	-0.08	0.00	0.00	0.18					
P194.9935+12.0585										
P198.0851+08.9428	0.10	-0.05	0.00	-0.01	0.11					
P207.3454+14.1268	0.09	-0.05	0.00	-0.03	0.11					
P207.9290+03.2767	0.12	-0.03	0.00	0.00	0.12					

Continued on next page

Table D.18 – *continued from previous page*

ID	[Na I/H]					[K I/H]				
	σ	ΔT	Δg	ΔFe	Δ	σ	ΔT	Δg	ΔFe	Δ
P209.0986+09.8244	0.07	-0.11	-0.01	-0.02	0.13					
P224.8444+02.3043	0.06	-0.04	0.05	0.00	0.09					
P236.9604+11.6155										
P237.8589+12.5660	0.18	-0.10	0.00	0.00	0.21					
P244.8986+10.9310	0.10	-0.09	0.00	-0.02	0.13					
P246.9682+08.5360						0.17	0.00	0.00	-0.05	0.18
P247.2115+06.6348	0.11	0.00	0.00	0.03	0.11					
P257.3131+12.8939	0.27	-0.05	-0.01	-0.03	0.28					
P258.1048+40.5405	0.16	-0.05	0.00	0.00	0.16					
P339.1417+25.5503										
P339.3203+25.8764	0.06	0.00	0.00	0.02	0.07					
HD 122563	0.06	-0.07	-0.02	0.04	0.10					

Table D.19: Systematic errors for Fe and Fe-peak elements. See the caption for Table D.16. Since only upper limits were determined for [Mn I/H], [Cu I/H], and [Zn I/H], for all stars in the sample, the systematic errors are not shown here.

ID	[Cr I/H]					[Fe I/H]					[Fe II/H]					[Ni I/H]				
	σ	ΔT	Δg	ΔFe	Δ	σ	ΔT	Δg	ΔFe	Δ	σ	ΔT	Δg	ΔFe	Δ	σ	ΔT	Δg	ΔFe	Δ
P008.5638+28.1855	0.11	0.00	0.00	0.04	0.11	0.05	-0.03	0.00	0.01	0.06	0.11	0.00	0.00	0.01	0.11					
P016.2907+28.3957	0.12	-0.03	0.00	-0.01	0.13	0.05	-0.03	-0.00	-0.00	0.06	0.08	0.00	-0.03	0.00	0.09	0.08	0.00	0.00	-0.02	0.09
P021.6938+29.0039	0.18	-0.10	0.00	0.01	0.21	0.05	-0.12	-0.01	-0.00	0.13	0.17	-0.02	-0.07	0.01	0.19	0.23	-0.10	0.00	0.01	0.25
P021.9576+32.4131	0.08	-0.02	0.00	-0.02	0.08	0.05	-0.02	0.00	0.00	0.05	0.08	-0.01	-0.02	0.01	0.08	0.11	-0.05	0.00	-0.02	0.12
P031.9938+27.7363	0.18	-0.05	0.00	-0.01	0.19	0.07	-0.06	-0.01	-0.00	0.09	0.10	-0.03	-0.03	-0.00	0.11					
P113.8240+45.1863	0.14	-0.02	0.00	0.00	0.14	0.05	-0.02	-0.00	-0.01	0.06	0.15	0.02	0.02	0.02	0.15					
P116.9657+33.5337	0.09	-0.02	0.00	0.01	0.09	0.04	-0.03	-0.00	-0.01	0.05	0.08	0.00	-0.02	0.01	0.09	0.11	-0.05	0.00	-0.04	0.13
P133.0683+28.7219	0.08	-0.03	0.00	-0.00	0.08	0.04	-0.02	0.00	-0.01	0.05	0.08	0.01	-0.03	0.00	0.08	0.27	-0.04	0.00	-0.01	0.27
P180.3206+02.5788	0.13	-0.05	0.00	0.00	0.14	0.05	-0.04	-0.01	-0.01	0.07	0.14	-0.00	-0.03	-0.00	0.14					
P182.5866+09.8940						0.06	-0.03	-0.01	-0.00	0.07	0.16	-0.02	-0.02	0.00	0.17					
P184.1783+01.0664						0.08	-0.23	-0.03	-0.04	0.25	0.18	-0.26	-0.22	-0.25	0.46					
P184.2997+43.4721	0.18	-0.05	0.00	-0.02	0.19	0.07	-0.02	0.03	0.01	0.08										
P188.0262+00.2055						0.06	-0.02	0.00	0.01	0.06	0.11	0.01	0.01	0.00	0.11					
P191.8535+12.0508						0.09	-0.05	-0.00	-0.01	0.11	0.12	-0.01	-0.01	0.01	0.12					
P192.3242+13.3956	0.14	-0.10	-0.02	-0.02	0.18	0.05	-0.09	-0.00	-0.01	0.10	0.11	-0.03	-0.04	-0.00	0.12	0.11	-0.05	0.00	0.00	0.12
P194.9935+12.0585						0.06	0.11	0.13	0.13	0.22	0.17	0.13	0.13	0.13	0.28					
P198.0851+08.9428	0.13	-0.07	0.03	-0.01	0.15	0.04	-0.09	-0.01	-0.03	0.11	0.14	0.05	-0.02	0.01	0.15	0.18	-0.06	0.00	0.00	0.19
P207.3454+14.1268						0.09	-0.03	0.00	-0.01	0.10	0.23	0.00	0.00	-0.01	0.23					
P207.9290+03.2767	0.13	-0.02	0.00	0.01	0.13	0.06	-0.03	-0.01	0.00	0.06	0.11	-0.00	-0.03	0.00	0.11					
P209.0986+09.8244						0.05	-0.15	-0.02	-0.01	0.16	0.10	-0.04	-0.04	-0.01	0.12	0.23	-0.10	0.00	0.04	0.25
P224.8444+02.3043	0.16	-0.05	0.05	0.00	0.17	0.05	-0.04	0.08	0.00	0.10	0.11	-0.00	-0.05	-0.02	0.12					
P236.9604+11.6155						0.10	-0.04	-0.00	-0.01	0.10	0.08	-0.01	-0.03	-0.01	0.08					
P237.8589+12.5660																				
P244.8986+10.9310						0.05	-0.11	-0.02	0.01	0.13	0.11	-0.03	-0.03	-0.01	0.12					
P246.9682+08.5360																				
P247.2115+06.6348	0.13	-0.02	0.00	0.00	0.13	0.05	-0.02	-0.00	-0.00	0.05	0.10	0.00	-0.03	-0.01	0.10	0.11	-0.05	0.00	-0.02	0.12
P257.3131+12.8939	0.16	-0.05	-0.02	0.01	0.17	0.04	-0.05	-0.00	-0.01	0.07	0.10	0.01	0.01	0.03	0.11	0.24	-0.05	0.00	-0.01	0.25
P258.1048+40.5405																				
P339.1417+25.5503	0.10	0.00	0.00	0.01	0.11	0.05	-0.04	0.00	-0.02	0.06	0.09	0.01	0.00	0.01	0.09					
P339.3203+25.8764	0.16	-0.05	-0.05	-0.03	0.17	0.06	-0.03	-0.01	0.00	0.07	0.09	0.02	-0.01	0.01	0.09					

Continued on next page

Table D.19 – *continued from previous page*

ID	[Cr I/H]					[Fe I/H]					[Fe II/H]					[Ni I/H]			
	σ	ΔT	Δg	ΔFe	Δ	σ	ΔT	Δg	ΔFe	Δ	σ	ΔT	Δg	ΔFe	Δ	σ	ΔT	Δg	ΔFe	Δ
HD 122563	0.08	-0.07	0.00	0.01	0.11	0.04	-0.07	-0.02	-0.01	0.09	0.06	0.02	-0.02	-0.01	0.07	0.05	-0.05	0.00	0.01	0.07

Table D.20 Systematic errors for [Ba II/H]. See the caption for Table D.16.

ID	[Ba II/H]				
	σ	ΔT	Δg	ΔFe	Δ
P008.5638+28.1855					
P016.2907+28.3957	0.11	-0.15	-0.17	-0.15	0.30
P021.6938+29.0039					
P021.9576+32.4131					
P031.9938+27.7363					
P113.8240+45.1863	0.12	-0.02	-0.02	0.00	0.13
P116.9657+33.5337					
P133.0683+28.7219	0.12	-0.05	-0.05	-0.02	0.14
P180.3206+02.5788					
P182.5866+09.8940					
P184.1783+01.0664	0.11	-0.12	-0.03	0.01	0.17
P184.2997+43.4721					
P188.0262+00.2055					
P191.8535+12.0508					
P192.3242+13.3956					
P194.9935+12.0585					
P198.0851+08.9428					
P207.3454+14.1268					
P207.9290+03.2767					
P209.0986+09.8244					
P224.8444+02.3043					
P236.9604+11.6155					
P237.8589+12.5660					
P244.8986+10.9310					
P246.9682+08.5360					
P247.2115+06.6348	0.13	-0.02	-0.02	-0.02	0.13
P257.3131+12.8939	0.12	-0.04	-0.02	-0.03	0.13
P258.1048+40.5405					
P339.1417+25.5503					
P339.3203+25.8764					
HD 122563	0.12	-0.05	-0.05	-0.02	0.14

Table D.21 Systematic errors for [Sc II/H], [La II/H] and [Nd II/H] for P016.2907+28.3957.

ID	[Sc II/H]					[La II/H]					[Nd II/H]				
	σ	ΔT	Δg	ΔFe	Δ	σ	ΔT	Δg	ΔFe	Δ	σ	ΔT	Δg	ΔFe	Δ
P016.2907+28.3957	0.35	0.00	-0.05	-0.02	0.36	0.20	0.05	0.05	0.03	0.22	0.20	0.06	0.06	0.09	0.24

Table D.22: NLTE Corrections. NLTE deviations were calculated on a line-by-line basis for each star given its line list and stellar parameters; the averaged NLTE correction is given below. O corrections from Sitnova et al. (2013), Na from Lind et al. (2011), Mg from Bergemann et al. (2017), Ca from Mashonkina et al. (2007), Ti from Bergemann (2011), and Cr from Bergemann & Cescutti (2010)

ID	Δ[O I/H]	Δ[Na I/H]	Δ[Mg I/H]	Δ[Ca I/H]	Δ[Ti I/H]	Δ[Ti II/H]	Δ[Cr I/H]
P016.2907+28.3957		-0.18 (2)	0.16 (4)	0.25 (6)	0.58 (2)	0.03 (6)	0.64 (4)
P021.6938+29.0039	-0.14 (1)	-0.08 (2)	0.11 (2)	0.27 (3)		0.05 (2)	0.56 (1)
P021.9576+32.4131		-0.15 (2)	0.18 (3)	0.28 (6)	0.60 (2)	0.06 (1)	0.67 (3)
P031.9938+27.7363	-0.16 (1)	-0.15 (1)	0.09 (3)	0.36 (3)		0.05 (2)	0.55 (1)
P113.8240+45.1863	-0.07 (2)	-0.41 (2)	0.02 (2)	0.08 (4)	0.34 (2)	0.03 (2)	0.36 (2)
P116.9657+33.5337		-0.23 (2)	0.13 (3)	0.24 (5)	0.61 (3)	0.05 (2)	0.59 (3)
P133.0683+28.7219		-0.26 (2)	0.13 (2)	0.25 (6)	0.62 (2)	0.04 (1)	0.58 (3)
P180.3206+02.5788		-0.12 (2)	0.15 (3)	0.24 (4)	0.59 (1)		0.65 (1)
P182.5866+09.8940		-0.08 (2)	0.07 (2)			0.05 (2)	0.47 (2)
P184.1783+01.0664		-0.27 (2)	0.16 (2)			0.05 (1)	0.62 (1)
P184.2997+43.4721		-0.10 (2)	0.01 (2)	0.14 (5)	0.27 (2)	0.04 (2)	0.45 (1)
P188.0262+00.2055		-0.08 (2)	0.07 (3)	0.17 (2)		0.05 (2)	0.47 (2)
P191.8535+12.0508			0.14 (2)				
P192.3242+13.3956	-0.09 (2)	-0.12 (2)	0.08 (3)	0.18 (4)	0.44 (2)	0.04 (1)	0.49 (2)
P194.9935+12.0585		-0.07 (2)	0.05 (2)	0.13 (2)	0.35 (1)	0.04 (2)	0.46 (3)
P198.0851+08.9428	-0.08 (2)	-0.22 (2)	0.08 (2)	0.28 (3)	0.46 (2)	0.04 (2)	0.70 (2)
P207.3454+14.1268		-0.17 (2)	0.10 (3)	0.22 (4)		0.04 (1)	0.50 (2)
P207.9290+03.2767	-0.21 (1)	-0.09 (2)	0.10 (2)	0.39 (2)		0.05 (2)	0.56 (2)

Continued on next page

Table D.22 – *continued from previous page*

ID	Δ[O I/H]	Δ[Na I/H]	Δ[Mg I/H]	Δ[Ca I/H]	Δ[Ti I/H]	Δ[Ti II/H]	Δ[Cr I/H]
P209.0986+09.8244	-0.21 (1)		0.11 (4)	0.39 (2)		0.05 (2)	0.58 (2)
P224.8444+02.3043		-0.09 (2)	0.19 (3)	0.25 (1)		0.00 (2)	0.70 (1)
P236.9604+11.6155		-0.26 (2)	0.23 (2)	0.28 (2)		0.05 (2)	0.75 (2)
P237.8589+12.5660			0.18 (2)				
P244.8986+10.9310	-0.29 (1)		0.12 (4)	0.42 (2)		0.05 (2)	0.59 (2)
P246.9682+08.5360		-0.08 (2)	0.17 (2)				0.63 (1)
P247.2115+06.6348		-0.15 (2)	0.12 (3)	0.28 (4)	0.61 (3)	0.09 (4)	0.59 (4)
P257.3131+12.8939		-0.26 (2)	0.14 (4)	0.22 (6)	0.62 (3)	0.07 (5)	0.58 (3)
P258.1048+40.5405		-0.07 (1)	0.13 (2)				
P339.1417+25.5503	-0.12 (1)	-0.10 (2)	0.06 (2)	0.20 (3)		0.05 (2)	0.46 (2)
P339.3203+25.8764		-0.12 (2)	0.15 (2)	0.20 (2)		0.04 (2)	0.64 (1)
HD 122563		-0.33 (2)	0.08 (3)	0.22 (7)	0.48 (4)	0.08 (7)	0.39 (5)

Table D.23 Sample line list of atomic lines used in the GRACES sample. The best fit abundance per line in each star is listed as $A(X) = \log[N(X)/N(H)] + 12$. Shortened target names are listed by their RA and DEC. Averaged abundances and line-to-line scatter or upper limits are reported in Table D.11.

Elem	λ (Å)	χ (eV)	$\log(gf)$	P016 +28	P237 +12	P247 +06	HD122563
Fe I	5269.537	0.858	-1.33	4.11	<3.77	3.94	4.36
Fe I	5328.039	0.914	-1.47	4.21	<3.97	3.99	4.36
Fe II	5169.028	2.891	-0.87	4.26	<3.47	4.19	
Li I	6707.924	0.0	-0.299	<0.81	1.92	<0.58	-0.97
O I	7771.944	9.139	0.32	7.22	<6.28	<6.77	
Na I	5889.951	0.0	0.12	3.10	2.36	2.84	3.42
Mg I	5172.684	2.71	-0.4	4.51	3.72	4.55	5.03
Mg I	5183.604	2.715	-0.18	4.66	3.87	4.50	5.13
K I	7698.974	0.0	-0.17	<2.49	<3.05	<2.23	
Ca I	6122.217	1.884	-0.41	3.75	<3.76	3.39	3.77
Ca I	6162.173	1.897	0.1	3.40	<3.36	3.19	3.47
Sc I	4670.4	1.356	-1.324	0.91	<2.07	<1.30	
Ti I	4981.731	0.848	0.57	2.36	<2.47	1.95	2.08
Ti I	4991.066	0.835	0.45	2.41	<2.47	2.00	2.13
Ti II	5188.68	1.581	-1.21	2.61	<2.47	2.15	2.58
Ti II	5226.543	1.565	-1.3	2.61	<2.47	1.95	2.43
Cr I	5206.023	0.941	0.02	2.40	<2.76	2.09	2.17
Cr I	5208.409	0.941	0.17	2.40	<2.66	1.99	2.22
Mn I	4823.514	2.317	-0.466	<2.54	<3.40	<2.53	2.21
Ni I	5476.904	1.825	-0.78	3.13	<3.34	2.92	3.20
Zn I	4722.153	4.027	-0.34	2.32	<3.28	<2.70	
Y II	4900.11	1.032	-0.09	-0.23	<0.68	<-0.52	-0.76
Ba II	6141.73	0.704	-0.077	1.09	<-0.40	-1.72	-1.59
Ba II	6496.91	0.604	-0.38	1.09	<-0.30	-1.72	-1.49
La II	4920.798	0.59	-0.27	-0.36	<0.62	<-0.80	<-1.72
Nd II	4811.342	0.064	-1.14	0.76	<2.04	<0.42	<-1.40
Nd II	4825.48	0.182	-0.42	0.06	<1.29	<-0.33	<-1.45
Eu II	6645.072	1.379	-0.517	<-0.52	<0.34	<-1.18	<-1.65

Table D.24: Orbital parameters and action vectors for stars in this paper, based on orbits calculated using the $1/\varpi$ distances. Actions and energies have been normalized to the Sun for brevity ($J_{\phi\odot} = 2009.92$ kpc km s^{-1}, $J_{z\odot} = 0.35$ kpc km s^{-1}, $E_\odot = -64943.61$ km^2 s^{-2}; Sestito et al. 2019)

ID	μ_α (μas yr^{-1})	μ_δ (μas yr^{-1})	r_{peri} (kpc)	r_{apo} (kpc)	Eccentricity	$|Z|_{max}$ (kpc)	$J_\phi/J_{\phi\odot}$	$J_z/J_{z\odot}$	$E/E_\odot$
P008.5638+28.1855	44.54 ± 0.07	12.19 ± 0.07	6.0 ± 0.1	11.7 ± 0.7	0.32 ± 0.02	10.9 ± 0.6	-0.27 ± 0.02	3113.98 ± 11.60	0.803 ± -0.028
P016.2907+28.3957	3.66 ± 0.06	-1.95 ± 0.05	5.7 ± 2.1	29.3 ± 7.0	0.68 ± 0.03	20.3 ± 1.4	-0.70 ± 0.40	2061.04 ± 572.95	0.391 ± -0.117
P021.6938+29.0039	13.42 ± 0.09	4.54 ± 0.07	4.8 ± 0.1	11.6 ± 0.9	0.42 ± 0.03	8.4 ± 1.6	0.45 ± 0.05	1532.21 ± 403.68	0.849 ± -0.038
P021.9576+32.4131	6.83 ± 0.09	-4.16 ± 0.08	3.3 ± 3.3	13.2 ± 3.1	0.67 ± 0.21	4.0 ± 1.2	-0.39 ± 0.48	254.75 ± 112.52	0.860 ± -0.159
P031.9938+27.7363	12.91 ± 0.09	-26.81 ± 0.10	2.5 ± 1.1	11.8 ± 0.6	0.66 ± 0.11	3.5 ± 1.4	-0.42 ± 0.12	307.35 ± 178.99	0.922 ± -0.062
P113.8240+45.1863	-33.61 ± 0.07	-23.27 ± 0.05	9.0 ± 0.1	20.9 ± 2.1	0.40 ± 0.04	17.9 ± 2.3	0.64 ± 0.02	3755.69 ± 437.76	0.496 ± -0.046
P116.9657+33.5337	0.94 ± 0.07	-11.58 ± 0.04	5.7 ± 3.9	45.4 ± 80.6	0.76 ± 0.09	24.0 ± 30.6	-0.77 ± 0.79	1440.86 ± 297.30	0.325 ± -0.292
P133.0683+28.7219	-0.73 ± 0.08	-11.92 ± 0.05	5.7 ± 4.9	16.4 ± 14.8	0.55 ± 0.25	6.9 ± 7.0	-0.61 ± 0.66	635.43 ± 383.09	0.748 ± -0.252
P180.3206+02.5788	-0.05 ± 0.10	-11.36 ± 0.06	4.4 ± 1.1	59.9 ± 78.7	0.82 ± 0.10	51.2 ± 58.1	0.25 ± 0.37	3365.47 ± 795.44	0.136 ± -0.228
P182.5866+09.8940	-76.32 ± 0.09	-46.52 ± 0.06	4.5 ± 0.4	34.2 ± 12.0	0.75 ± 0.05	10.5 ± 4.6	-0.87 ± 0.10	372.56 ± 89.22	0.339 ± -0.162
P184.1783+01.0664	-1.92 ± 0.12	-6.88 ± 0.07	8.3 ± 0.1	27.3 ± 0.8	0.53 ± 0.01	26.2 ± 0.8	-0.30 ± 0.12	5894.07 ± 637.49	0.383 ± -0.014
P184.2997+43.4721	-19.30 ± 0.05	-37.97 ± 0.06	1.8 ± 0.1	8.5 ± 0.0	0.64 ± 0.02	2.5 ± 0.2	0.31 ± 0.02	245.74 ± 4.87	1.140 ± -0.005
P188.0262+00.2055	25.18 ± 0.10	-35.02 ± 0.07	1.6 ± 0.2	21.4 ± 1.9	0.86 ± 0.02	4.1 ± 0.4	0.34 ± 0.03	145.12 ± 5.97	0.600 ± -0.051
P191.8535+12.0508	-11.78 ± 0.08	-0.00 ± 0.05	4.7 ± 0.2	9.5 ± 0.2	0.34 ± 0.03	2.5 ± 0.2	0.64 ± 0.02	233.92 ± 21.55	0.958 ± -0.003
P192.3242+13.3956	-12.93 ± 0.09	-1.36 ± 0.06	2.6 ± 0.4	11.1 ± 0.8	0.62 ± 0.07	4.2 ± 0.7	0.42 ± 0.05	392.90 ± 52.80	0.946 ± -0.031
P194.9935+12.0585	-32.89 ± 0.07	-27.50 ± 0.05	0.6 ± 0.2	8.2 ± 0.4	0.86 ± 0.04	3.8 ± 0.4	0.10 ± 0.03	349.95 ± 37.17	1.183 ± -0.004
P198.0851+08.9428	-0.22 ± 0.10	-0.72 ± 0.07	23.9 ± 28.5	86.0 ± 247.6	0.33 ± 0.20	55.6 ± 85.7	-0.06 ± 0.56	15444.69 ± 23639.44	0.204 ± -0.445
P207.3454+14.1268	-6.03 ± 0.14	-16.95 ± 0.10	2.6 ± 0.6	8.0 ± 0.1	0.52 ± 0.09	3.6 ± 0.4	0.35 ± 0.08	485.66 ± 64.42	1.124 ± -0.021
P207.9290+03.2767	-27.35 ± 0.07	-9.93 ± 0.05	0.9 ± 0.4	11.6 ± 65.9	0.83 ± 0.07	8.4 ± 59.7	-0.12 ± 0.07	813.71 ± 58.63	1.065 ± -0.043
P209.0986+09.8244	-20.90 ± 0.08	-1.71 ± 0.06	1.5 ± 0.4	14.1 ± 2.0	0.80 ± 0.07	10.4 ± 3.3	0.19 ± 0.08	1086.27 ± 316.95	0.825 ± -0.073
P224.8444+02.3043	-2.31 ± 0.06	-0.10 ± 0.06	13.4 ± 13.6	331.6 ± 662.9	0.76 ± 0.15	239.9 ± 476.0	0.57 ± 1.10	11107.29 ± 13163.12	-0.105 ± -0.728
P236.9604+11.6155	-0.86 ± 0.09	-3.12 ± 0.14	4.1 ± 0.6	6.3 ± 0.3	0.21 ± 0.06	3.9 ± 1.0	0.38 ± 0.11	883.84 ± 371.10	1.147 ± -0.035
P237.8589+12.5660	-5.94 ± 0.06	-7.86 ± 0.05	2.1 ± 0.4	6.6 ± 0.1	0.52 ± 0.07	2.5 ± 0.4	0.30 ± 0.06	328.14 ± 71.82	1.268 ± -0.026
P244.8986+10.9310	-16.05 ± 0.06	-12.30 ± 0.04	1.1 ± 0.6	7.9 ± 14.5	0.74 ± 0.14	2.3 ± 6.0	-0.17 ± 0.09	209.31 ± 17.05	1.291 ± -0.013
P246.9682+08.5360	-6.41 ± 0.05	12.37 ± 0.04	3.7 ± 0.4	28.7 ± 6.8	0.76 ± 0.07	9.5 ± 3.5	0.69 ± 0.04	398.12 ± 84.91	0.424 ± -0.118
P247.2115+06.6348	-1.27 ± 0.06	-2.96 ± 0.03	2.6 ± 0.9	7.9 ± 2.6	0.47 ± 0.21	7.2 ± 1.9	-0.02 ± 0.15	1900.73 ± 456.17	1.101 ± -0.116
P257.3131+12.8939	2.14 ± 0.07	-1.62 ± 0.06	8.4 ± 3.5	23.5 ± 32.4	0.37 ± 0.12	22.7 ± 32.3	-0.06 ± 0.26	5781.40 ± 3366.22	0.610 ± -0.363
P258.1048+40.5405	0.83 ± 0.06	1.07 ± 0.08	3.0 ± 0.0	8.3 ± 0.0	0.47 ± 0.00	4.7 ± 0.2	0.37 ± 0.01	736.29 ± 55.41	1.076 ± -0.005
P339.1417+25.5503	33.41 ± 0.06	11.41 ± 0.05	2.6 ± 0.1	12.8 ± 0.4	0.66 ± 0.02	0.8 ± 0.0	0.50 ± 0.01	22.68 ± 0.72	0.883 ± -0.014

Continued on next page

Table D.24 – *continued from previous page*

ID	μ_α (μas yr^{-1})	μ_δ (μas yr^{-1})	r_{peri} (kpc)	r_{apo} (kpc)	Eccentricity	$|Z|_{max}$ (kpc)	$J_\phi/J_{\phi\odot}$	$J_z/J_{z\odot}$	$E/E_\odot$
P339.3203+25.8764	20.30 ± 0.08	2.16 ± 0.06	2.0 ± 1.1	46.0 ± 76.7	0.90 ± 0.02	9.4 ± 19.7	-0.45 ± 0.22	157.15 ± 57.13	0.337 ± -0.308